AF802982

Anu Hopia and Erik Fooladi

A Pinch of Culinary Science

Boiling an Egg Inside Out and Other Kitchen Tales

CRC Press
Taylor & Francis Group
Boca Raton London New York

CRC Press is an imprint of the
Taylor & Francis Group, an **informa** business

Copyright page

CRC Press
Taylor & Francis Group
6000 Broken Sound Parkway NW, Suite 300
Boca Raton, FL 33487-2742

© 2019 by Taylor & Francis Group, LLC
© Book design, layout, illustration, graphic and information design by Aki Scharin

CRC Press is an imprint of Taylor & Francis Group, an Informa business

No claim to original U.S. Government works

Printed on acid-free paper

International Standard Book Number-13: 978-1-138-61130-6 (Hardback)

This book contains information obtained from authentic and highly regarded sources. Reasonable efforts have been made to publish reliable data and information, but the author and publisher cannot assume responsibility for the validity of all materials or the consequences of their use. The authors and publishers have attempted to trace the copyright holders of all material reproduced in this publication and apologize to copyright holders if permission to publish in this form has not been obtained. If any copyright material has not been acknowledged please write and let us know so we may rectify in any future reprint.

Except as permitted under U.S. Copyright Law, no part of this book may be reprinted, reproduced, transmitted, or utilized in any form by any electronic, mechanical, or other means, now known or hereafter invented, including photocopying, microfilming, and recording,or in any information storage or retrieval system, without written permission from the publishers.

For permission to photocopy or use material electronically from this work, please access www.copyright.com (http://www.copyright.com/) or contact the Copyright Clearance Center, Inc. (CCC), 222 Rosewood Drive, Danvers, MA 01923, 978-750-8400. CCC is a not-for-profit organization that provides licenses and registration for a variety of users. For organizations that have been granted a photocopy license by the CCC, a separate system of payment has been arranged.

Trademark Notice: Product or corporate names may be trademarks or registered trademarks, and are used only for identification and explanation without intent to infringe.

Library of Congress Cataloging-in-Publication Data

Names: Hopia, Anu Inkeri, author. | Fooladi, Erik Cyrus, author.
Title: A pinch of culinary science : boiling an egg inside out and other
kitchen tales / Anu Inkeri Hopia, Erik Cyrus Fooladi.
Description: Boca Raton : CRC Press, Taylor & Francis Group, 2019. | Includes
bibliographical references.
Identifiers: LCCN 2018051394| ISBN 9781138611306 (hardback : alk. paper) |
ISBN 9780429878442 (pdf) | ISBN 9780429878435 (epub) | ISBN 9780429878428
(mobi/kindle)
Subjects: LCSH: Cooking--Popular works. | Food habits--Popular works. |
LCGFT: Cookbooks.
Classification: LCC TX652 .H6477 2019 | DDC 394.1/2--dc23
LC record available at https://lccn.loc.gov/2018051394

Visit the Taylor & Francis Web site at
http://www.taylorandfrancis.com

and the CRC Press Web site at
http://www.crcpress.com

Contents

Preface

For a natural scientist, food is usually molecules, plant cells, and phases of matter such as liquids, solids, and gases. Chemical reactions in the kitchen dictate the result, physical phenomena such as heat play important roles and the whole thing ends up as a meal and an experience.

No surprise that we, two food-loving chemists, would choose such an entrance to the book we write. Who we are would surely put its stamp on the way we talk about the food we cook and eat. But you don't necessarily need to have an academic degree in chemistry to harness these chemical substances and reactions. This is clearly visible every day as skillful chefs, professional or amateur, apply their experience-based knowledge handed down from one person to another all over the world. While we are professionally excited about all the scientific knowledge that chemistry offers to us, we are aware that food is much more than the sum of its molecules. Food is also a part of our common and personal history, woven into the identity of every human being. It is part of our culture, of our great and small emotions and experiences. With food, we communicate, we make friendships, and we show love and affection. We hope this can be seen throughout the book as it would be sad if the food was reduced to molecules and salts—mere outcomes of chemical reactions and the laws of physics.

Often, the most commonplace transformations in the kitchen have prompted us to ask why things happen as they do, and this has been the starting point of most chapters in this book. In the context of our open and informal scientific food workshops, arranged in both Finland and Norway, we have approached various questions from both theoretical and practical points of view with the aim of understanding food, meals, and joys thereof. We have not only read books and scientific publications, but we have also carried out experiments together in the attempt to reach the very heart of the phenomena. Thus, we do not want to just convey funny anecdotes or fun facts about food chemistry or to pretend that we know exactly what's going on at the molecular level in food. Some things we know, but much we do not know, and we are curious and will never stop asking questions. Therefore, we try to ask questions about issues that have not already been described in the research literature, can be found in cookbooks, or by a search on the Internet. And often when we look for the answer to a question, we find only part of an answer, and, in addition, two new questions appear. Rather than giving unambiguous answers, we would like to invite the reader to consider these kitchen mysteries with us. This way we hope to pass on curiosity about food, sow seeds for the reader's own inquiries in the kitchen, and promote diversity in food and food culture.

Two other persons have been important along this journey. Chef Tatu Lehtovaara, who has shared his expertise and experience in the art and craft of the kitchen, has been involved in creating and implementing the food workshops from which most of the examples in the book are taken. Most of the book's recipes are by Tatu's hand, and he has also raised a significant number of interesting questions that we ourselves wouldn't have envisaged but have been shown to be both significant and thought provoking.

The other important person in the book is graphic artist Aki Scharin, who has dazzled us with an imaginative journey into the world of food structures and molecules. Although Aki's illustrations usually set out from food and molecules, often, at some point, they begin to live their own lives. Through his work, the food and its molecules develop to become part of our common visual heritage. As a result of this mutual effort, this book of cooperation and friendship has emerged.

Helsinki and Volda
Anu Hopia and Erik Fooladi

Acknowledgments

We have already thanked chef Tatu Lehtovaara and graphic artist Aki Scharin, who both have played an important role in making this book into what it has become. Tatu has been a driving force in the startup and implementation of the food workshops in Helsinki; he has contributed to the planning of every workshop. With Aki we have been working for almost ten years. Even without any science background, he has a unique ability to visualize fundamental natural and scientific phenomena, and his references and connections move in the most unlikely and unthinkable directions. The many discussions over illustrations have been exceptionally stimulating, probably for all three of us. His skills and imagination both amaze and delight.

Many others are also to be thanked for their contributions. A special thanks to all you who have participated in, and hosted, the food workshops in Finland and Norway. For your interest and active participation, for contributing with your knowledge and experience and for giving us such interesting questions to explore. This should be clearly visible for those reading the book.

We have received help and input from many while writing the various chapters. We would like to thank chef Arto Rastas (65°C egg recipe), Maria Planting (meringue discussions), Ismo Korhonen (discussions on steaming mussels as well as cooking for and conducting the porridge tasting), Christophe Lavelle (discussions on heat), Guro H. Rognså (white wine discussions), Gunnar Aurdal (music descriptions), Liv Marit Mathilde Aurdal (discussions on archeology and prehistoric cooking), Pasta Factory Helsinki Oy (specially made pasta for a workshop), Nanna Rintala (translation of Finnish excerpts from the flavor chapter into English), Pekka Lehtonen (knowledge on wine), Øyvind Andersen (dishwashing discussions) and the three informants Liv Brekke, Margit Fet and Jostein Fet (traditions related to porridge).

The workshops described in the book have been conducted in various places in and around Helsinki, and in the Sunnmøre region in Western Norway: Stadi Vocational School (Helsinki, Finland), Klippfiskakademiet (Ålesund, Norway), Borgund Upper Secondary School (Ålesund), Heureka Finnish Science Center (Vantaa, Finland), Alko Oy Training Center (Helsinki), Teurastamo Kellohalli (Helsinki), E. Ahlström Showroom (Helsinki) and Unilever Suomi's test kitchen (Helsinki). Thank you very much, all of you—it has been a wonderful journey and we hope the cooperation may continue.

Our respective families have experienced both suffering and enjoyment as a result of our work with this book. Thank you for always being there!

Anu Hopia and Erik Fooladi

1

The Real Thing? The Chemistry and Psychology of an Apple Pie

On a dark and rainy October afternoon, a group of our gastro-scientific food workshop participants gathered to enjoy some apple pie, and to discuss the role of molecules in our food. The Chef and the Chemist had each baked a delicious apple pie for the event and the participants were asked to decide and tell which they preferred and why. Both apple pies were baked with the same type of commercially available pie crust, the difference was in the fillings. While the Chef made his pie filling by sautéing Granny Smith apples in butter and sugar, the Chemist made hers trying to keep the best possible control of which molecules went into the pie: she added carefully weighed amounts of specific molecules to a filling that had never been in the vicinity of an apple. The apple-like taste in the Chemist's mock apple was created using the main chemical substances that makes apple taste like apple: sugar and fruit acids. A structure similar to apple compote used in many apple pies was achieved using salted crackers soaked in the sweet-sour liquid.

Mock apple pie

Ingredients

Ready-made pie dough
500 ml water
250 g sugar
2.5 g malic acid
2.5 g tartaric acid (these fruit acids are available in home-brewing and wine stores and some supermarkets)
30 round salted crackers (e.g. Ritz)
Cinnamon
Butter or margarine

Procedure

1. Mix the water, sugar and acids and heat to a boil.
2. Place the salted crackers carefully into the boiled mixture and heat gently for about 3 minutes until the crackers have swollen and absorbed the liquid to the full. Keep the crackers under the liquid but do not stir, thus avoiding breaking the crackers. When swollen they will gain a structure and mouthfeel resembling apple compote.
3. Press the pie dough into a 24–26 cm round baking tin.
4. Use a sieve to let excess liquid drip off the crackers and move the "compote" carefully onto the pie dough (the pie easily becomes too moist).
5. Sprinkle with cinnamon and add some lumps of butter/margarine.
6. Bake the pie in a medium hot oven (225°C) for approximately 30 minutes until the crust and filling have developed a nice, golden brown color.

The two apple pie twins were cooked in the same oven and then served for blind tasting by the participants, unknowing of the fact that one was made with apples and the other without. The participants were asked to describe the two pies, and then report which pie they would prefer. The nine participants constituting the evening's taste panel were, however, not only asked about which they preferred, but also but also *why* they did so. Not surprisingly, the pie made of real apples disappeared from the plates rapidly and was the one preferred by the majority: the votes were seven for the Granny Smith version and two for the "mock" version. The majority reported that the Chef's Granny Smith pie had a better structure and stronger apple flavor, and was simply "better tasting." The two choosing the Chemist's pie acknowledged this to have better appearance and nice moistness. This rather depressing result for the Chemist made us wonder whether the mock apple pie could be improved to compete with the "real" one, or whether the latter would always be preferred regardless of cooking skills and recipe. After all, the result might be a matter of cooking skills rather than recipe.

It was surprising that none of the participants questioned the nature of the Chemist's mock apple pie. The majority preferred the "true" pie, but they still accepted the apple-free pie to be an apple pie. All the important apple pie characteristics were present: sweetness from the sugar, crisp sourness from the fruit acids, and nice and moist apple compote-like structure from the softened crackers. Even many of the characteristic elements of apple pie aroma were there: cinnamon and the caramelized flavor from heated sugar and butter. For obvious reasons, the apple flavor was rather weak: the fruit acids are not volatile and do not contribute any apple aroma. What if we had added a drop of some of the organic esters that are major aroma compounds found in apples?

All in all, after thorough discussion, the participants considered the mock apple pie to be a decent alternative in case you get visitors allergic to apple or if you have forgotten to buy apples (and by chance had crackers, malic acid, and tartaric acid lying around). But if you ask us, the fact that it is possible to make a mock apple pie is motivation enough in itself to make the attempt.

What makes an apple pie authentic? Cooking with pure molecules does sound like a very modern, and even cold, issue and the discussion very often ends up around food additives and unwanted substances in our food. However, apple pie without apples is not a new invention. On the contrary, numerous mock apple pie recipes from the late 1800s are found in old cookbooks, and by a quick internet search. In these old recipes, mock apple pies were prepared at purest using dry crackers, sugar and tartaric acid to create an apple illusion. Sometimes apple or lemon juice was available and could be used to enhance the flavor. After all, even during times of scarcity, people still longed for familiar comfort foods and maybe an apple pie would lighten up life even more than in times of abundance. Housewives were quite innovative indeed, but how little apple can you accept in order to call it an apple pie? Where do we draw the line? The mock apple pie prepared "molecule by molecule" was not a true apple pie, because, according to common food legislation, you are not allowed to mislead the consumer by incorrect wording. As a manufacturer you wouldn't be allowed to sell the 19th century, or great-grandma's, mock apple pie under the name of "apple pie."

The psychology of eating a mock apple pie. What do you feel about the notion of eating an "apple pie" where the apple flavor is represented by a mixture of malic acid and tartaric acid? Or what if it was made using 2-hydroxybutanedioic acid and 2,3-dihydroxysuccinic acid? Would it in a sense be "wrong," making "artificial food" in our very own kitchen similar to those fruit flavored sweets and candies that can be bought in shops and supermarkets? Undoubtedly, these issues stir the emotions in many of us as we, through media, read and hear that we should rather make "pure food" from scratch, avoiding what is often termed "processed food" and foods containing many additives.

Moving into the landscape of emotions, we turn to the psychologists rather than chemists to help us understand what is going on in our minds. In 2009, University of Michigan researchers Hyunjin Song and Norbert Schwartz published an intriguing research article that sheds light on the relation between additives and emotions. They asked respondents to read ten names of food additives and rate them according to the experienced degree of risk associated with eating them, from very safe to very harmful. The results showed that respondents rated the names that were *hard to pronounce* as more hazardous than those with easy-to-pronounce names. In the same study, the researchers asked respondents to rate names of roller coasters according to which they thought would make them feel most sick. In this case, they arrived at the exact same result: if the name was difficult to pronounce, the respondents felt the rollercoaster would make them feel sicker. Consequently, the researchers named their article, "If It's Difficult to Pronounce, It Must Be Risky." Words that were difficult to pronounce were associated not only with "risky," but also with "new." In our apple pie context, if we use the two names from above, 2-hydroxybutanedioic acid and 2,3-dihydroxysuccinic acid, when listing the ingredients, some might feel less well eating it, even though these names are simply the formal chemical names for malic acid and tartaric acid. After all, malic acid is the major acid in apples and tartaric acid the most predominant acid in grapes (in all Nordic languages, the two compounds are indeed called the equivalent of "wine acid" and "apple acid").

There are numerous books about additives, taking various perspectives and views. We will not go into the discussion for or against additives, except acknowledging that they are among the most thoroughly tested ingredients we eat. They are often necessary in order for our food to be safe, have the required texture and so forth, and many of them are even sourced directly from the world of plants with little processing. The famous food writer and documentarist Michael Pollan is often quoted on his statement, "Don't eat anything your great-grandmother wouldn't recognize as food." However, in the case of the mock apple pie, it is indeed our great-grandmother who baked the pie based on flavors from additives rather than apples! Surely, there is no rule without a few exceptions. The fact that the participants at the workshop didn't appear very skeptical to the mock apple pie might be because they were already well seasoned through participation in many such science-based food workshops, or they might have been familiar with chemistry and, therefore, accustomed to chemical language. If you wanted to convince an "additive skeptic" to eat the mock apple pie, perhaps the knowledge that it is cooked from an old "traditional" recipe would balance the feeling of risk and novelty and instill a notion that it is safe to eat? As a contrast to the additive skeptic, the French researcher Hervé This has, together with food professionals, spent much effort in developing dishes, and even whole meals, based solely on combinations and manipulation of pure chemical substances, that is, "note by note cooking." That must be the ultimate way of cooking from scratch. When, or if, such foods are made available for a broader community, the producers will surely have to reflect upon which words they use in their ingredient lists.

In addition to our knowledge of the ingredients' names, the name of a dish might affect our perception of it. Researchers Line Holler Mielby and Michael Bom Frøst from the University of Copenhagen found that when (very fortunate) volunteers were served an avant-garde fine dining meal that included some "challenging" courses, the manner in which the waiter described the dish affected how the guests experienced the food. The guests gave a more positive evaluation of a course if the waiter presented it with descriptions of the ingredients and the procedure of cooking and assembling the dish. The same course was rated lower if it was presented accompanied by an artful name, or simply without any comment at all. This is something to reflect upon the next time you have someone for dinner that you expect will be skeptical about what you serve them. Indeed, language has great power over our experiences of food and drink!

2

Boiling an Egg from the Inside Out

In 2006, we got our hands on a book we had waited for with great anticipation, written by the French professor Hervé This, a pioneer within molecular gastronomy since its very conception. The book, *Molecular Gastronomy*, published in French four years earlier, was finally available in English translation. On a side note, during our years of waiting, NASA had (in 2003–04) found time and resources to launch and land two rovers on the planet Mars. More on this later in the chapter. We were especially intrigued by what we read in a chapter about cooking eggs. It says: "At 62°C one of the proteins in the white (ovotransferrin) is cooked, but the yolk remains liquid because the proteins that coagulate first in this part of the egg require a temperature of 68°C. Obviously this would mean longer cooking times, but the result is a perfectly cooked egg" (see table with egg white proteins in the next chapter). So, the egg yolk and white contain their particular cocktails of proteins, coagulating at different temperatures. In everyday language this would mean that the egg white becomes solid at 62°C and the yolk solidifies at 68°C. Eureka! Imagine the classical hotel breakfast egg problem: you pick an egg from the basket promising soft-boiled eggs, but when you open the egg you realize it would have survived a round of boccia without taking too much damage. We had at last been served the solution to this dilemma. Fill a pot with water, make sure the temperature is between 62°C and 68°C, pop in a few eggs and wait. After a while, although longer than if the egg was cooked in boiling water, you would get a risk-free soft-boiled egg. This knowledge could change the way millions of eggs are cooked all over the world. Hotels could now guarantee perfect soft-boiled eggs from 7am until the breakfast buffet closes around 11am. The chefs can basically forget the cooking time, just make sure they have strict control over the temperature.

To the lab! We didn't waste much time and went straight to the lab next morning. We set up a thermostat-controlled water bath at exactly 65°C and popped in four eggs. But for how long should the eggs cook? We chose to be on the safe side and went to the office for the rest of the day, returning to the lab in the afternoon. Protein chemistry has taught us that each protein has its specific conditions, such as temperature and pH under which it denatures, its molecular shape being changed and finally coagulates into a solid substance. In the protein chemistry books there is usually no mention about reaction time however. Thus, two of the eggs were extracted after a short day's

Proteins are long molecules, composed of hundreds or thousands of amino acids linked together in a long chain, which is again bundled together in a specific way. The shapes of the bundles are affected in varying degrees during cooking. Subjected to either heat, physical stress (as in whisking egg whites), or in contact with acids, the bundles start unfolding. *Denaturing* is a mild structural change which is not always visible to the eye; the egg white is still translucent and the meat of a steak has not changed its color or texture markedly. Harsher treatment of the proteins will result in *coagulation*, a stronger and irreversible change in protein structure. The strings start reacting and hook up with each other to form a three-dimensional network. This is clearly visible to the naked eye: an egg white changes from translucent to opaque, meat or fish becomes firm, scrambled eggs become lumpy, milk curdles. ×

work, and the remaining were left to cook until next day. As long as the temperature was held below 68°C, the yolk should not coagulate, right? Giving two eggs the same treatment would ensure us some measure of repeatability; at least, if the same thing happened with both eggs. In case they came out different, we'd know that we should repeat the experiment.

The first two eggs were cracked open after six hours in the bath. There, on the plate, a whitish, slushy and rather liquid white poured out while the yolk rolled out in the shape of a soft ball. Not only one egg, but both. It seemed as if the eggs were cooked from the inside out! Next morning, 26 hours after they entered the 65°C water, the remaining two eggs were extracted. The observation was more or less the same as for the first two eggs, but the yolks were even more firm while the white was still not completely set. During the subsequent weeks and months, the experiment was repeated several times for different cooking times and temperatures between 62 and below 68°C, but the results were more or less the same. How could this happen? Why didn't we get the expected perfectly cooked eggs according to the literature? Our yolks had become solid already at 65°C, while the whites were still not properly set, which seemed to be very different from what we had read about in the book. We searched scientific and popular scientific literature to find a straight explanation, but in vain. After all, most people cook their eggs in boiling water, close to 100°C. Could it be that no one had actually tried cooking eggs at other temperatures? Or maybe someone had, but hadn't been among those writing research reports or cookbooks about it?

The physicists' eggs. When cooking eggs, the traditional way, at around 100°C, we rely on the laws of physics. The water is boiling hot, the egg is cold and the heat would expectedly spread from the outside in. Since all parts of the egg coagulate well below this temperature, the egg will solidify from the outside in. The white will become solid before the yolk is heated enough to become firm, and the higher the temperature, the firmer the texture. With this tactic, your best tool for a successful result is a timer. If you want to be very accurate, add a tape measure. Because larger egg means longer cooking time. This way of cooking eggs inspires physicists, in their spare time, to create complicated mathematical formulas for egg cooking, in turn ending up as egg cooking calculators on the web. Advanced versions take into consideration the desired outcome (soft, medium, hard yolk), the size of the egg, starting temperature of the egg (did you take it straight from the fridge or was it kept in room temperature?) and even elevation above sea level. The starting temperature of the egg makes a difference; if you start with a room-temperature egg, it will be as much as 16–20°C warmer than taken straight from the fridge, and, not surprisingly, this will require longer cooking times. The Norwegian egg cooking calculator tells us that taking a medium-sized egg straight from the fridge requires about 40 seconds longer cooking time compared with one kept at room temperature when you want it cooked to medium soft yolk. But why does the calculator include elevation above sea level? The boiling point of water depends on the surrounding air pressure, which changes with elevation, in turn affecting the cooking time. High up in the mountains at 2000 meters, water boils at just below 98°C. For a small-to-medium-sized egg this amounts to 30 seconds difference in cooking time for a medium-soft yolk. Such calculators, or equations, assume that you drop the egg straight into boiling water. If you start with tap water, pop in the egg and then turn on the heat, the calculator does not work because the water temperature will be constantly changing the first minutes. Making a functioning equation or calculator for cooking the egg from cold water at home would be a virtually impossible task because of the many factors that come into play in different kitchens around the world: the size and material of the pan, the amount of water, the type of hotplate and so forth. It seems that the world has not yet seen such a passionate soft-boiled-egg-loving physicist as such an equation and calculator has not been developed thus far.

Cooking eggs in cooling water. Some rely on a trick that they claim will ensure a good result every time: heat water to the boil, pop in the eggs, take the pan off the hotplate and leave for a certain amount of time. As a general principle, this is actually a very old cooking technique described elsewhere in the book, dating back to prehistoric times: the cooking pit. In this case, hot rocks are used instead of water. This way of cooking the eggs relies on the ability of water to store the energy it receives from the hotplate, and subsequently to give it away as thermal energy to the food. The longer the food is cooked, the cooler the water becomes, so the risk of overcooking the eggs is not as high as when keeping the eggs at constant boiling temperature of 100°C. The same principle is also evidenced when cooking porridge, where the milk and rice is heated to the boil, wrapped in newspapers or a duvet and left to cook by the residual heat for a few hours. However, inherent in this method are some major pitfalls as the end results are very dependent on how much heat is stored in the system: do you have one liter of water or 10 liters? What material is the pan made of? Is there heat stored in your conventional cast iron hotplate, or maybe you use a convection hotplate of ceramic material that doesn't store that much energy? If you heat one liter of water to 100°C, pop in an egg, and leave it for 15 minutes, you will most likely get a very different result compared with heating 100 liters of water to the same temperature still cooking one egg. One hundred liters of water will stay at a higher temperature for a longer time than one liter. And popping one egg into one liter of water should give a rather different result than cooking five at a time, because the five eggs will cool the water more than one egg, and they will have to "share the energy" from the same amount of hot water as compared to the single egg. Add to this, cooking eggs where the margins between undercooked, perfect and overcooked are narrow. The only way to assess the state of matters is to crack the egg open: the ultimate point of no return. So, if a cookbook was to give a recipe for cooking eggs in cooling water, it has to specify not only temperatures and time, but, at the least, also the amount of water per egg. But for your own kitchen, when you use the same procedure, the same amount of water and number of eggs every time, it might work out perfectly fine every time. You might even achieve some credibility among friends and family for being able to cook perfect eggs, according to your taste, in a very consistent manner. But it might be risky to share your procedure with others unless you give fairly detailed instructions.

Cooking eggs at 6X°C - relying on chemistry. So, what happened with the eggs? What was the reason for our "Eureka!" moment turning into a typical "What the...?" moment even though we had been very careful controlling the temperature? The answer did not arrive until 2011, in a scientific paper by César Vega and Ruben Mercadé-Prieto called "Culinary Biophysics: on the Nature of the 6X°C

Egg." Our inside-out-cooked eggs had now got a name of their own: the 6X°C egg. The experiment that the two researchers did was quite simple, and very clever. They heated a large number of 18g samples of egg yolk to respective temperatures between 60 and 66°C for different lengths of time. Then they measured the viscosity, how soft or hard the yolk had become, for each combination of time and temperature. What they found was that the texture of the yolk was not only a question of temperature, but also of time, even at stable, low temperatures. In their research paper, the two authors presented this in a graph showing how the texture developed at various temperatures (see diagram). At last, these strangely cooked eggs had been described by science, or at least vital information—a missing piece of the puzzle—for us to understand how eggs behave had eventually surfaced.

The viscosities, or textures, of the yolks were given as numerical values with the unit Pa·S. The researchers must have wished that their results should not be relevant only to the research community

Effect of cooking time and temperature on the structure of the yolk

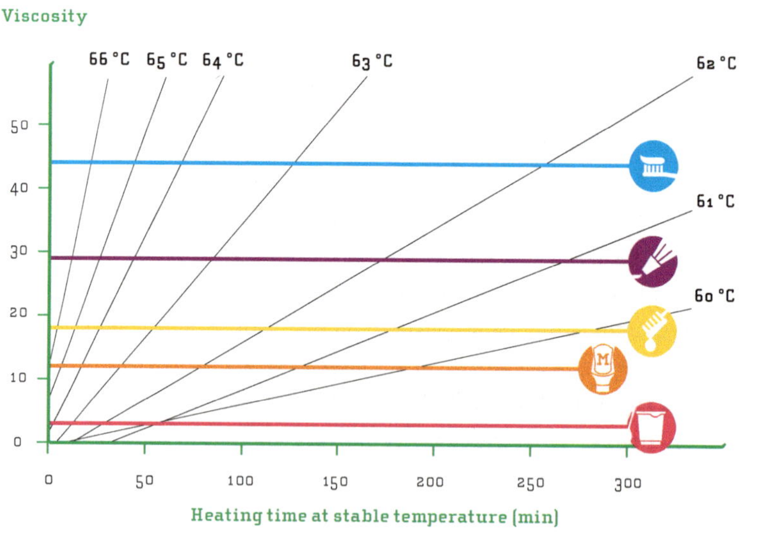

^ During cooking, the structure of the yolk depends on both time and temperature. Pa·S is a numerical unit for viscosity. The higher the value, the more viscous (thicker) the yolk. The illustration is based on the original diagram by Vega and Mercadé-Prieto (2011).

Viscosity of selected foods
(adapted from Vega and Mercadé-Prieto, 2011)

Food	Viscosity (Pa·s)
Whipping cream	0.02
Raw egg yolk	0.09
Pancake syrup	0.96
Chocolate syrup	1.4
Sour cream (17% fat)	2.9
Greek-style yogurt	3.0
Molasses	3.3
Sweetened condensed milk	6.8
Mayonnaise	12.1
Ready-to-eat chocolate pudding	13.8
Honey (liquid honey)	18.3
Nutella®	28.1
Cookie icing (fresh)	29.3
Toothpaste	43.8
Marmite®	43.9

^ The table shows viscosity of various foods. High value signifies high viscosity. The table is adapted from Vega and Mercadé-Prieto (2011).

in soft matter physics and the like, but also for the rest of us. As a thoughtful gesture, they included a table where they linked viscosity values with familiar foods. After all, you wouldn't ask the waiter that you would "like to have your egg cooked to 12 Pa·s at 10 s^{-1}." On the other hand, asking for the yolk to have similar texture as mayonnaise would give meaning to most people. Reading off the graph, a texture similar to mayonnaise would require cooking the eggs in 64°C for ca. 15 minutes, 63°C for 40 minutes, or 62°C for 75–80 minutes.

The question, however, remains: Is there any point in making things more intricate than they are? After all, you might say that your Sunday egg comes out satisfactorily almost every time, that this is good enough for you, and that you would rather spend your energy on the Sunday newspaper. There are indeed reasons to try it at least once. For one, the texture of a 6X°C egg is quite different than what you can ever get by cooking the egg at 100°C. It is difficult to explain in a few words, so you just have to try it. Secondly, maybe we can return this question with another question: why put robots on Mars? Why should curiosity to inquire be limited to space scientists, rocket engineers, and the like? Space research is often justified by the fact that many inventions come out of work that was never meant to produce that specific invention. In the case of the 6X°C egg, this happened almost instantaneously. As soon as this way of cooking eggs became known, creative chefs started inventing new dishes, bringing our thoughts to a quote from the famous French gastronome Jean Anthelme Brillat-Savarin (1755–1826): "The discovery of a new dish does more for the happiness of the human race than the discovery of a star." Brillat-Savarin is also the writer famous for saying, "Tell me what you eat, and I will tell you who you are." It is worth reflecting upon that in the seven-year period in human history, between 2004 and 2011 while we could boast about our ability to put a functioning remote-controlled vehicle on Mars, we didn't have the scientific knowledge of what goes on inside the millions of eggs cooked in kitchens around the world every day. On the other hand, maybe there was a chef, a housewife, or someone else loving their Sunday egg that knew about the 6X°C egg phenomenon all this time without sharing it with the rest of us through research journals or cookbooks?

Visualization of temperature development in eggs cooked at 100 and 64°C, respectively, for 8 and 45 minutes.

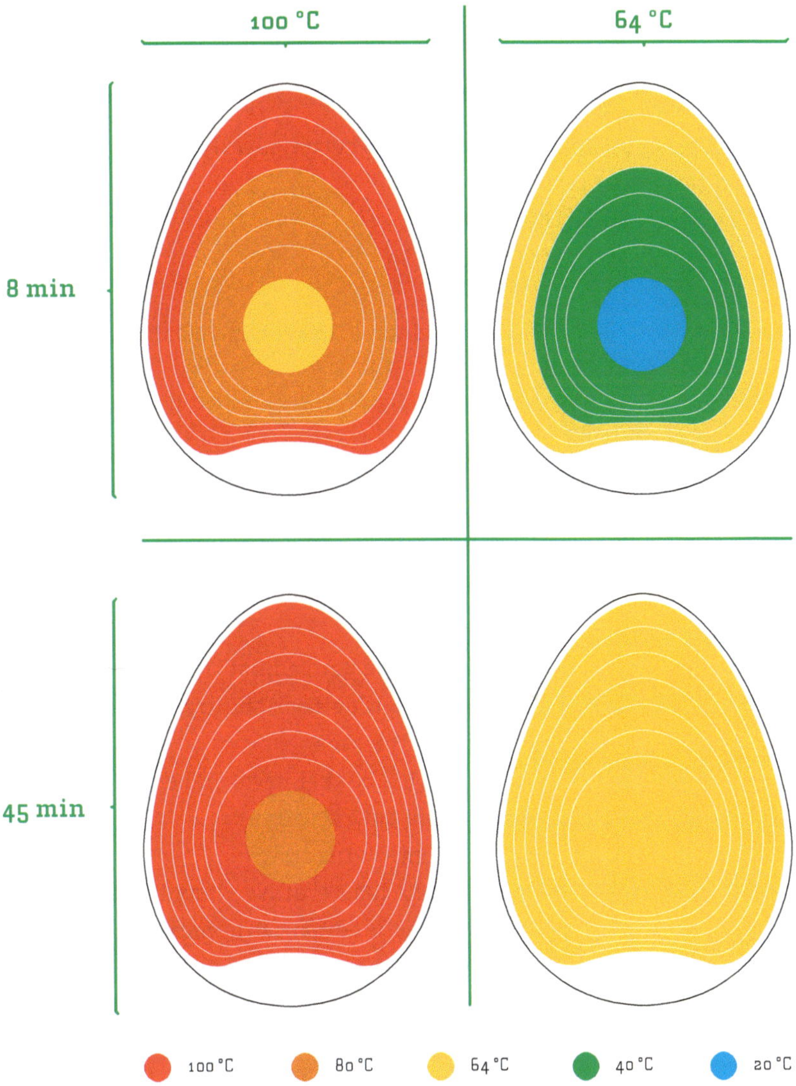

DIY procedure for cooking 6X°C eggs

If you have a thermostat-controlled water bath in your home, this is easy. Since most of us don't, we give a basic low-tech strategy that works fine.

Equipment and materials

Large pan with lid (the larger the better, ideally at least 5 l)
Water cooker or small pan
Digital cooking thermometer
Water
Eggs
(Recommended but not mandatory: basket or a mesh bag for the eggs)

Procedure

1. Start heating ca. 1 l water to the boil in the little pan or water cooker.
2. Fill the large pan ¾-full with water as hot as possible straight from the tap. Put the pan on the hotplate on low/lowest power setting.
3. Put the thermometer in the large pan, adjust the water temperature to desired temperature with either boiling or cold water. Leave the thermometer in all the time.
4. If available, place the eggs in a basket or mesh bag to prevent them touching the hot bottom part of the pan.
5. Pop in the egg(s) and make note of the time.
6. Put the lid on and monitor the temperature along the way. You may need to scoop out water to adjust with more hot or cold water. Eventually, you'll get to know your hotplate. Make sure to stir the water occasionally to avoid temperature variations between lower and higher layers of water.
7. Leave the eggs in for the desired amount of time.

Tips

- Suggested temperature and time for first attempt: 64°C for one hour.
- Small temperature changes might make a difference, so make sure to check the temperature and stir the water frequently.
- Why not turn the hotplate on max to begin with? Because (a) there is an imminent risk of temperature gradients in the pan, warmer at the bottom where the eggs are and cold at the top by the thermometer; (b) it is much more difficult to maintain a stable temperature.

Serving suggestions

To find the most recent innovations, make an internet search for "6X°C eggs" or "sous vide eggs." Alternatively, you might want to try Anu and chef Arto Rastas' recipe for the "deconstructed fried egg:"it is much more difficult to maintain a stable temperature.

Crack each ready-cooked 6X°C egg into its respective little bowl. With gentle hands, take out the yolk and clean the white off, transferring the yolk to a pan of lukewarm oil with some chopped chives. Fry each of the whites in some oil or butter/margarine at medium temperature, ideally using a small circular mold. Arrange the white on a plate, put the yolk on top and sprinkle with flake salt and freshly ground black pepper. If you want to make many deconstructed eggs, use an oiled silicone or metal muffin pan and bake the whites in the oven at medium temperature until firm. Alternatively, you can cook and serve the eggs directly from small ramekins (tall creme brûlée-type). In that case, bake the whites in a water bath in the oven in order to give them a more gentle treatment.

3

In the Beginning There Was an Egg

How would our culinary universe look if we did not have the egg? No boiled, fried, poached or scrambled. No omelets. How would you make an angel cake or muffin if you didn't have an egg? We would live in a cake-less world. Not even pancakes. Just think about all the pancakes in the world—they would all be non-existing without eggs. How would you make sauces such as mayonnaise, Hollandaise or sabayons without eggs? Custard for your pies, or delicate meringue desserts such as pavlovas or baked Alaskas? Indeed, egg is a cornerstone of our gastronomy without which the most important foams, gels, and emulsions would be practically impossible to make. Or would they really?

Meringue from egg white or pea? The fact that you can make a perfect meringue completely without egg whites is not an invention that should be credited to any chemist. On the contrary, chemists have kept on telling the story about the wonderful proteins of egg whites—how the different types of molecules cooperate to decrease the surface tension of water, increase its viscosity, and finally gently denature onto the surface of air bubbles to form a thick and elastic film to produce airy, strong and long-lasting foams. This foamy structure is in turn used for culinary purposes in many dishes

^ **A good stiff foam is mostly air.** Small pockets of air are trapped in thin layers of water with a mixture of molecules suitable for stabilizing a foam. The illustration is based on a microscope picture of egg white foam taken during a workshop.

The role of egg white proteins in cooking

Protein	%	Denaturation temperature/°C	Role in foaming
Ovalbumin	54	84.5	Denatures during whisking
Conalbumin/ ovotransferrin	12	61.5	Does not denature during whisking
Ovomucoid	11	70	Strong structure, does not denature during whisking
Ovoglobulins	8	92.5	Lowers surface tension, increases viscosity
Lysozyme	3.4	75	Lowers surface tension, increases viscosity
Ovomucin	3.5	70	Forms denatured film on surface of bubbles
Other proteins	Ca. 8		

^ Various proteins, various roles. Source: Belitz, Grosch and Schieberle (2009).

The powerful foam of chickpea liquid. We wanted to study if the rumor was true—could the chickpeas' foaming capacity compete with that of egg whites? We bought three boxes of cooked chickpeas (unsalted, organic) and randomly took three eggs from the fridge. The latter move would later show itself as a mistake. We will return to that issue later. The workshop attendants were divided into two groups: the foam group and the meringue group. Both groups were given two KitchenAid machines, identical sets of other necessary kitchen utensils and instruction for the speed setting for the KitchenAid machine. The first group was asked to report on the time it took to produce a foam, foam volumes right after whisking and how long it took for the foam to start to leak liquid. The second group was asked to compare the two foams in with respect to aroma, taste, and texture.

The first group noted that the egg white at 12.5 minutes was somewhat quicker to form a foam compared with the chickpea liquid at 13.5 minutes. But the chickpea juice gave a slightly higher volume (2.7 dl vs. 2.2 dl for the egg white), and the former gave a far more stable foam than the egg white. Indeed, while the egg white foam started leaking after only 5 minutes, the chickpea foam only started leaking after 15 minutes. So that evening we learned that chickpea juice can indeed be a very potential foaming agent! Making foam was just as easy for both; chickpeas produced more foam for the same amount of liquid, and it was even more stable.

But the weakness of our egg white foam made us suspicious of the quality of our eggs, so we measured the pH of the whites with a pH indicator paper. Our suspicion was confirmed: a pH of 9.5! In fresh eggs, the pH of the white is commonly around 6, from where it rises as the egg ages. What an embarrassing mistake! While attempting to adhere to the scientific principle of picking eggs randomly we had picked old ones! Such a high pH indicates that the eggs are really quite old and will therefore have lowered foaming properties and

During storage, carbon dioxide gradually seeps out through the egg shell. Dissolved in water, carbon dioxide is in the form of the weak acid carbonic acid. As long as the carbonic acid is present, the pH in the egg is kept in check, close to neutral. But as the carbon dioxide evaporates, the pH starts to rise and can finally reach values as high as 9.5—alkaline indeed, and a clear sign of mature age. Aging also breaks down the proteins responsible for producing and stabilizing foam. The egg white solution also is less viscous due to the gradual breakdown of the long protein molecules. The result is egg whites that are difficult to whisk to a foam, a foam that collapses more easily and a liquid that is more easily drained off the bubbles due to gravity drains, ultimately giving a dry and brittle foam with a puddle of watery liquid at the bottom of the bowl. ×

poor water-holding capacity. So, this comparison was probably quite unfair on behalf of the egg white. But still, we had shown that chickpeas are a potent competitor, and at least in a battle with old eggs, chickpea juice was the winner.

The second group who tasted the two (uncooked) meringues noted that they had very similar flavor and texture properties, indeed. There were some differences, however. The egg white version appeared to be shinier and the foam denser and smoother than the chickpea version. The mouthfeel of the chickpea version was claimed to be thinner and it also melted quicker in the mouth. The taste of the chickpea version was somewhat sweeter and the aroma also differed slightly in the two versions. There was a noticeable beany flavor in the chickpea meringue. But this was surprisingly weak, and the group speculated that it might fade away during cooking. The citrus flavor was somewhat more present in the egg meringue, while the chickpea version had a slightly more pronounced sweetness. All in all, the egg white version seemed to get a slightly more favorable evaluation, but with a narrow margin. The tasting was not a blind tasting so we cannot disregard possible preconceptions among those who tasted the two sweet foams to have affected the result. Considering that most of the panelists prior to the test had indicated their doubts about the power of the chickpea, this meringue came out very favorable. On top of this, the beany flavor of a chickpea meringue might be suppressed when the meringues are baked, so we might even have done the chickpeas some injustice.

In 2013, food scientist Johanna Söderberg submitted her PhD thesis at the Swedish University of Agricultural Sciences about foaming, emulsifying and gel forming properties of legume-derived proteins, as alternatives to those derived from eggs. Indeed, it seems that one can make vegan, foamy delicacies such as meringue, or emulsions such as mayonnaise, by replacing egg with soybean or pea proteins as long as they can be dissolved in water. Canned peas and beans are normally made by cooking them at approximately 120°C under high pressure. During this process, proteins, and possibly also other foaming agents such as carbohydrates, are dissolved in the cooking liquid, which thereby turns into an excellent foaming agent. It is not only chickpeas that contain significant amounts of proteins with similar properties to those in eggs. Various blogs and other open sources give long lists of ingredients suitable for this purpose: white beans, red beans, lentils, soybeans, gluten, rice, hemp, and quinoa. This is surely an invitation to start experimenting! In the future, we should expect to be offered foods in which egg or other animal proteins are partly or wholly replaced by what many of us thought should be discarded, but recently has received its own name as a functional ingredient: "aquafaba." Literally meaning bean water (aqua = water, faba = beans), the word received its place in the *Oxford English Dictionary* in 2016. Johanna Söderberg's results indicate that we probably have to accept a mild beany flavor if we want to use these vegan alternatives. But she also observed that as much as 25% of the egg proteins could be replaced by their legume-derived counterparts without any noticeable effect on flavor or texture.

Even though food scientists studied legume constituents for decades, apparently few had thought that the world of vegetables had an ingredient that could compete with egg whites as the basic structure for foams in our home cooking. But then some food writers showed us that it is no big deal to make wonderful vegan foams. Actually, we are more than happy to have been shown up. Indeed, everyone deserves to be able to enjoy the fluffy happiness of meringues, soufflés and other airy delicacies! And surely, science will follow with research to back up the experiments of bloggers, chefs, and others, be it professionals or amateurs, sharing their experience with the rest of us.

4

Mussels on Acid

We were sitting at a small Norwegian seafood restaurant with pints of beer and a pot of mussels in front of us. Life couldn't have been better: the company was good, beer gently foaming and the hot steaming mussels better than ever. Indeed, we were convinced that the mussels that we had this night, cooked in local beer, were more tender and moist than their comrades that we had tasted before, who had met their fate in steaming white wine. Usually, observations made over a pint of beer are not considered the hallmark of scientific validity. However, even after a good night's sleep the impression persisted of having had unusually succulent and tender mussels the night before. Surely, they were freshly caught from the Atlantic, the convivial atmosphere and nice people around the table had added to the overall experience. The question was thus: would our observation at the restaurant stand up to a replication study under more neutral circumstances? There are obviously a number of possible factors behind our informal observation, so we had to narrow it down to one single research question: the effect of using beer when steaming mussels. A question of chemical nature. How do beer and wine differ as liquids for steaming mussels, and what effect might they have on the final product? The first thing coming to mind was the difference in acidity. Wine is more acidic than beer as the pH of beer usually ranges between 4 and 4.5 whereas the pH of wines is lower, between 2.9 and 3.9. Hence, wines ranger from slightly to considerably more acidic than beers, and it is well known that pH has an effect on both chemical reactions during cooking and the structure of food.

Blind tasting of mussels

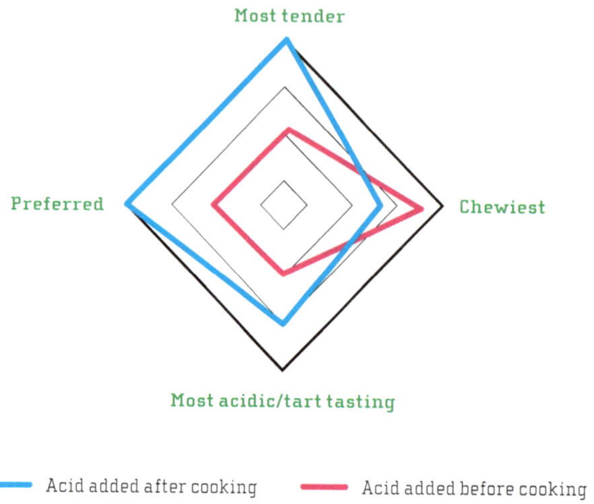

^ The diagram shows how the panel of 29 participants judged the mussels. One portion was steamed with citric acid, and the other had citric acid added after steaming.

Which recommendations should we give based on this experiment? At the very least, if you want to cook your mussels in white wine, you might want to pick a wine with moderate acidity. But beer or good fish stock may also be good alternatives.

Acids in cooking. Protein chemistry plays a key role when it comes to structural changes of food during cooking. When protein-rich foods such as fish, meat, or egg are exposed to heat, the natural structure of proteins is disturbed. The protein molecules denature and eventually coagulate, resulting in changes such as firmer structure and changes in appearance. In addition to heat, there are at least three other strategies to changing the protein structure in food: we can use acids to "cook" protein-rich foods. An example is the Peruvian dish ceviche, where the cook sprinkles lime juice over raw fish cut in small pieces, or the Nordic pickled herring, where herring or Baltic herring is "cooked" in vinegar, often with added spices and herbs. The acid in lime juice or vinegar affect the structure of the fish meat: the almost transparent and soft structure turns opaque and takes a firmer "cooked" structure. The second strategy is cooking with ethanol, or, in everyday language, alcohol. In some recipes for carpaccio, which is based on thin slices of raw meat, we are asked to add a dash of brandy or other strong spirit. Doing the same with salmon, and adding some salt and sugar, as well, we get gravlax. The spirit does not only add flavor, but the alcohol also gives a more tender meat because high concentrations of ethanol make proteins denature. The third strategy is described in the chapter on foams, such as when we converted the clear fluid of egg white into an opaque and stiff foam by sheer physical movement of a whisk. The change of appearance from clear to opaque indicates that we have actually "cooked" the egg white proteins by our vigorous whisking, spreading and stretching the egg white protein bundles around the surface of air bubbles.

Combine any of these treatments and you get new kinds of solid structures because you expose the raw material to more than one treatment. What we observed in our mussel experiment can also be seen when you poach an egg. Recipes for poached eggs recommend a little vinegar in the cooking water. The egg white coagulates faster and you get a characteristic firmer structure on the outer layer of the egg white. Without vinegar you simply get a more classic soft-boiled egg. Thus, playing with acid and combining it with heat or other methods that affect protein structure adds diversity to the culinary structures of everyday food. But be aware that combining such strategies can also result in overcooking your mussels, such as using a too acidic wine or not reducing the heat treatment when combining it with acid.

Acids can also result in less pleasant surprises when cooking. We experienced this when baking a traditional Finnish spice commonly made for Christmas. The cake batter contains a special mixture of gingerbread spices as well as a significant amount of lingonberries, a tart and somewhat bitter wild berry commonly used in the Nordic countries.

Spice cake with lingonberries

Ingredients

100 g butter or margarine
2 dl brown sugar
3 eggs
2.5 dl flour
2 tsp baking soda
1 dl mashed lingonberries
1–2 tsp cinnamon
1–2 tsp ginger
0.5 tsp cloves

Procedure

1. Whisk butter/margarine and sugar into a dense foam.
2. Whisk the eggs into the butter/sugar foam one by one.
3. Mix baking soda and spices into the flour and fold it gently into the foam.
4. Add lingonberries and transfer to a 1.25 l springform pan.
5. Bake at 175°C for 45 minutes or until the cake is ready (poke a knitting needle or wooden cocktail stick into it. If the batter doesn't cling to the stick, it should be ready).

The unexpected happened when we, in a hurry, added the ingredients in a different order than instructed by the recipe; the flour and lingonberries changed order. Rather than beating sugar and butter with eggs, adding flour and finally lingonberries, we beat sugar and butter with eggs, added lingonberries, and stirred in the flour at the end. The acidic lingonberries made a significant part of the egg proteins, which were supposed to maintain a foamy and spongy structure, denature already in the cold batter—well before the cake went into the oven. This coagulation should not take place until after the cake is in the oven. The foam is a delicate structure, and the batter must first be allowed to rise (by effect of the leavening agent and water evaporating), and thereafter solidify into a firm but porous cake. But since some of the egg proteins had already coagulated in the batter due to the acids from the berries, there was not much flexible protein network left to form a structure around the bubbles in the foam. The hard work of beating and whisking air into the sugar, butter, and eggs had been in vain. Ready baked, the cake came out as a sad, flat, and crumbly specimen. Damn acids! Luckily the cake was served to a chemist colleague, so one positive side effect of the accident was the pleasure of a discussion about protein chemistry.

5

Sausage Engineering

How many sausage recipes could there be in the world? Bratwurst, chorizo, kiełbasa biała, frankfurter, chipolata, salami, blood sausage, seitan sausage, baloney. The sausage is part of culinary heritages all over the world, and has since prehistoric times been a method to prepare and preserve meat, making use of all parts of the animal. Salted meat, offal, or blood is mixed with spices and perhaps some cereals, stuffed into casings made from various natural materials such as stomach, lungs or intestines of cattle, pigs, or sheep. The resulting neat packages can be dried, smoked, fermented, cooked or a combination of these treatments, to be enjoyed at a later point. Even after new storage techniques were invented such as ice houses, fridges, and freezers, the sausage was not abandoned—perhaps because it is such a practical technique to deal with meat, and the end result so delicious. On the contrary, the idea of sausage has been carried on and developed further by modern home and professional butchers, and later in large scale by the food industry. And the fact that more and more sausages based on vegetarian protein sources are making their way to the consumers is another sign that the popularity of sausages seems to increase rather than decrease.

No doubt that the sausage is an important part of the world's food heritage. For example, in the United States, the National Hot Dog and Sausage Council reported that only during summer, the three-month-long peak hot dog season, 320 million American citizens typically consume 7 billion hot dogs. Around 20 hot dogs per person. Also, Europeans and those in Nordic countries love sausages. The Finns eat approximately 20 kg sausages per capita a year, which is about half a barbeque sausage every single day for every Finn on earth. Although Norwegians are not the winners in this somewhat peculiar sausage consumption competition, they are reported to consume around 115 sausages per capita annually, one sausage every third day. Of course, many are not sausage fans, so if we remove these from the statistics it is obvious that those who eat them have one or two per day, around the year. Consequently, which type of sausages we make or eat is no small issue, and worthy of closer inspection.

Mixtures of things that do not mix. Substances that make up food come in various forms, or states (phases) of matter: solid, liquid or gas (plus a couple of other, rather rare ones of little relevance to home cooking).

Food commonly consist of a mixture of substances in various states, and more often than not it is a mixture of substances of rather different natures that do not normally mix. Oil and water are two liquids that will not dissolve into each other but remain separate even after vigorous mixing, returning to separate layers shortly after you stop mixing. To get these two mutually opposing materials to form a fairly stable mixture, an intermediary is required: an emulsifier. The emulsifier does not make the oil dissolve in water, but makes sure they are kept together; we have an emulsion. Emulsifying agents are usually molecules with both water-loving (hydrophilic) and fat-loving (lipophilic) properties in different parts of one and the same molecule. Examples of such emulsifiers are lecithin found in, for example, egg yolks and soybeans, phospholipids that make up the cell membranes of cells, certain proteins and many more.

What about making your sausages at home? Although most of us let the food industry make our sausages, stuffing one's own sausages at home has become increasingly popular. It is not particularly difficult and can be done with relatively simple means. All you need is meat (or some plant protein such as seitan), casings and grinding and a stuffing attachment for your kitchen machine. All of it is available from the shelf or fridge in any well-stocked supermarket and a kitchen equipment shop.

Regardless of type, sausages are all complex mixtures where fat and water form a delicate and mutual relationship. The key ingredients of a typical sausage are meat, fat, and water. Making good sausages, be it in industry or at home, is based on quality raw materials and the building of a stable emulsion where fat droplets are dispersed in a water-based continuous phase (see info box and diagram). In this mixture, the basis is a water-based, protein-rich mixture into which fat is emulsified into oil droplets or solid fat particles. Some sausages have a coarse structure with noticeable bits of meat and fat, such as salami or the Spanish-cured salchichón. However, if we want smooth and juicy sausages, the fat droplets need to be really small in size to give that creamy sensation on the tongue rather than clumps of fat rolling around in the mouth. Furthermore, small fat droplets are better retained in the sausage during cooking, rather than leaking into the pan, taking with it the good flavor and juicy texture into the water. Too little fat will give hard, crumbly and dry sausages. In order to get a juicy and smooth texture it is recommended to use at least 20% fat.

In sausage mince, the most important emulsifying agents are the meat proteins, with both water soluble and fat-soluble properties. They make sure that the fat particles don't separate out. In addition to the casings you will at least need meat/protein, fat, salt, and water to make a good sausage. All of these four play their own important roles in achieving a good texture. If one is missing, you will probably fail, or you might struggle convincing others that you have made a sausage. The bulk of the sausage is meat, which provides the important proteins that emulsify the fat particles as well as water. In some recipes the water bound up in the meat is not sufficient, and you are asked to add some additional water. Upon cooking, the proteins coagulate to form a firm but nicely elastic structure in the ready-cooked sausage. Salt helps dissolve some of the proteins in the water which in turn helps the fat becoming evenly dispersed and emulsified in the mass/dough.

Mayonnaise is an emulsion classic, where small oil droplets are surrounded by lecithin from the egg yolk and evenly distributed in water from lemon juice or vinegar. In that case, the oil is the dispersed phase while the water is the continuous phase, both still being liquid. We have a water-in-oil (w/o) emulsion. Milk and cream are emulsions of the same kind, where either oil droplets or fat particles (depending on the temperature) are dispersed in water. The same applies to butter and margarine, but in those cases water droplets are dispersed in oil or fat: an oil-in-water (o/w) emulsion. Many such mixtures consisting of otherwise non-miscible or difficult-to-solve substance pairs exist (see the diagram). Examples are suspensions (solid particles floating in a liquid), foams (gas bubbles trapped in a solid or liquid) and gels (liquid bound in a solid network). When whisking double cream, three difficult-to-mix materials dance together in an airy whipped cream while the whisk gives the rhythm: air, water, and fat. A general term for such mixtures of immiscible substances is dispersions. Indeed, we should all praise the chef who can organize mutually repellent molecules into a wide range of different structures for our mouth to enjoy!

But why introduce all these new and foreign words? Well, because they give us a language that allows us to talk about food as general structures, and thereby make sense of the food from a structural perspective. A chef can, for example, use the term "emulsion sauce" as a general term for sauces with certain properties (Béarnaise, Hollandaise and so forth). The language that follows with this is then a tool that helps us in understanding and dealing with them in practice, such as understanding why the sauce occasionally splits, how to keep it stable and so forth. ×

The structures of foods and other mixed materials

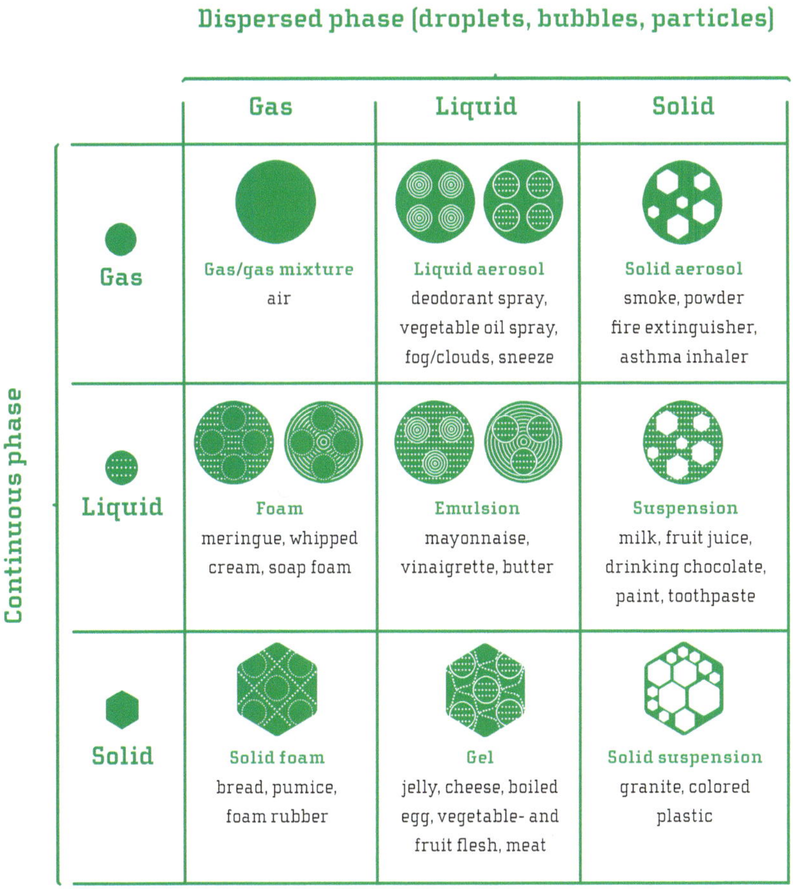

^ **Diagram with possible dispersions.** Both components of a dispersion (the dispersed and continuous phase) can be either gas, liquid, or solid, giving nine possible combinations. The gas mixture in the upper left-hand corner is the only exception, not being a dispersion but one single homogeneous phase. Thus, there exist eight basic dispersions.

^ **Sausages are dispersions:** The most important components are water (blue, continuous phase), fat (yellow, dispersed phase), meat proteins (red) and connective tissue proteins (green). The meat proteins surround the fat and function as an emulsifier. The two types of proteins permeate the mixture and makes a sausage both an emulsion and a gel.

The Food and Agriculture Organization of United Nations, FAO, have much information on their web sites, including detailed descriptions of what happens during the various steps in sausage making:

1. During grinding the meat, muscle- and connective tissue fragments are mixed to form a homogeneous mass.
2. Addition of salt and water causes meat fragments to swell. Furthermore, part of the proteins will dissolve in the salt water to form networks in the liquid—a gel. Other parts of the muscle proteins remain intact and fairly unaffected, floating in the water to form a suspension.
3. Fat is then added, resulting in an emulsion in which the complex water-meat mixture described above forms the continuous phase and fat taking role as dispersed phase.
4. During heating the protein molecules go through structural change to coagulate and firm up to form a rigid network: The soft and pliable sausage mass turns into cooked sausage. The connective tissue fragments mainly consist of the protein collagen with a triple helix structure. Upon heating, this untangles into single strands of gelatin proteins which dissolve in the hot water, eventually to form a gelatin gel when the sausage is later cooled (see figure). The more connective tissue in the meat, the higher more gelatin in the cooked sausage.

The process gives a product that consists of several states of matter and various structures at the same time, giving the texture and mouthfeel we have learnt to enjoy. A sausage is at the same time an emulsion, a suspension, and a gel. Sometimes there are air bubbles or -pockets in the structure, so foam can also be said to occur in some sausages. Evidently, foods are often not only dispersions, but mixtures of dispersions.

Sausage on the rocks. In some sausage recipes, especially from old cookbooks, a rather specific and peculiar advice is given. Namely that the water should be added in the form of ice. In more modern recipes this is not usually mentioned, but the advice to keep the ingredients cold during the mixing is not uncommon. Why not just use water from the tap? This is one of those questions that begs for an experiment to be done during one of our scientific food workshops. Why ice? Why cold? Is it simply because the raw materials are easily perishable, or could there be another reason?

We prepared two batches of basic pork sausages with only four ingredients: meat, fat, salt, and water. The sausage batches were identical except for the water temperature. In one batch water was added as crushed ice while the other batch got room-tempered water. The outcome was then blind tasted by the nine participants who were to evaluate, without mustard or any other condiments, which sausage was most juicy, which had the smoothest texture and which was most salty. Naturally we also asked them to point out which of the two they would prefer. We also collected data of somewhat "harder" nature: the weight of the samples before and after cooking, and the temperature of the sausage mass at the end of mixing.

The temperature measurements showed that ice cooled the sausage mass effectively, with a temperature difference of more than 10°C between the sausage masses after mixing (5.5°C vs. 16.9°C). Visually the samples appeared to be similar, and this difference in temperature during preparation seemed to be the only difference between the two sausage batches. However, during cooking the two behaved rather differently: starting with 160 g samples, we saw a clear weight difference between the final, steamed, sausages. The sausages with ice had lost about 5 g while the sausages with room-tempered water had lost more than double of the amount. The sausages also tasted differently, as two-thirds of the participants considered the ice versions to have smoother texture. Eight out of nine considered the sausages prepared with room-tempered water as saltier, and some said that the saltier sausages gave a stronger burst of salty water when biting into the sausage. When it comes to juiciness and preference, the two performed equally, or the difference was too small to discern.

The conclusion from our informal experiment was that the water temperature indeed can make a difference when you make sausages. The weight and texture differences may be due to a stronger emulsion/gel and thus better water-holding capacity of the meat proteins in the colder mass. A more stable emulsion and gel may also explain the difference in saltiness and burst of salty juices, since a weaker dispersion will more easily release water.

Sausage On The Rocks

Ingredients per batch

600 g finely minced pork meat
190 minced pork fat (lard)
125 g water (as crushed ice or room-tempered)
16.5 g salt (1.8 %)

Procedure for the two batches

Room-tempered water

1. Meat and fat were mixed in a kitchen machine using the K-beater.
2. After two minutes water was added (22°C), and mixing continued until a total of 15 minutes.
3. The temperature was measured (16.9°C).

Water as crushed ice

1. Meat and fat were mixed in a kitchen machine using the K-beater.
2. After two minutes crushed ice was added (0°C), and mixing continued until a total of 15 minutes.
3. The temperature was measured (5.5°C).

For both batches

4. Four portions of each 160 g was rolled into 36 mm diameter sausages shaped cylinders using plastic wrap.
5. The sausages were coded so the taste panel should not know which was from which batch. They were then cooked in steam oven at 80°C for 22 minutes.
6. The sausages were weighed.

Results of cooking loss measurements and sensory evaluation of sausages.

Average weight/g	Juiciest	Smoothest	Saltiest	Preferred
Recipe with crushed ice				
156 g +/- 2 g	5	6	1	4
Recipe with room-tempered water				
147 g +/- 2 g	4	3	8	5

^ The table shows the participants' verdict in a blind tasting of sausages made with water added as crushed ice and room-tempered water, respectively. Numbers refer to the number of votes; the starting weight for all sausages was 160 g.

The temperature issue is also mentioned in the technological sausage literature, where it is recommended that the temperature of the dough/mass should be kept below 12°C during mixing. There are two common reasons given to back up this recommendation: manufacturing hygiene and cooking loss. Our results thus were in accordance with the observations made in the food industry: lower dough temperature gave lower cooking loss.

However, why one should keep the temperature low to minimize cooking loss has two competing explanations in the literature, of which one or both might be correct. The first is about overheating. During grinding and mixing, "hot spots" with considerably higher temperature than the rest of the mass might occur. Indeed, one of the most prominent proteins in meat, myosin, denatures at 40°C. And if the protein is denatured, it will have reduced capacity to form a water-holding gel or stabilize the fat-in-water emulsion. But if the temperature is kept low, the risk of such premature protein denaturation due to "hot spots" is minimized. The second explanation concerns the fat: the harder fat (cold fat is harder than warm fat) will result in smaller fat particles. Thereby, the fat is distributed more evenly throughout the mass, and the emulsion is more stable. A more stable emulsion does not so easily leak fat nor water.

Which of the explanations is the most plausible, or which reason has the greater effect on the final product, is still a matter of debate among meat scientists. However, there is no problem agreeing on the fact that temperature has an effect on the final quality of sausages. In the discussions following our experiment, we agreed that this debate among sausage scientists is of utmost importance and that there is indeed need for more research on this area. So, for now, we will have to live with the fact that we know that this works, but not exactly why. We cross our fingers that the sausage scientists will work hard and do quality research in the near future.

6

Mustard—Fiercely Yours

The cups and the pots and the pans are to the kitchen what beakers, flasks and test tubes are to the chemistry laboratory. Every time you cook, dozens, hundreds or even thousands of reactions take place, more or less simultaneously. Some of these reactions are by nature enzymatic, where enzymes play a vital role either before, during or even after cooking. Enzymes are a subgroup among proteins that function as catalysts, the worker molecules among proteins that facilitate various reactions that otherwise would not have occurred or that would have required much heating. Since many enzymes in plant and animal cells are still operational in non-heated food materials, vegetables, fruits, meat, and even the wheat flour that we buy from grocery stores are full of different types of active enzymes.

A very familiar enzymatic reaction to most of us is the formation of strong, pungent substances from garlic and onion that can make us cry. Whole onion and garlic are very mild in flavor with no detectable pungency or kick. The pungency emerges if the plant cells are crushed and enzymes come in contact with certain compounds elsewhere in the cell. The unharmed cell has an organized structure where enzymes and their target molecules are isolated from each other, but when we start cutting, bruising, or chewing the onion, at the same time wreaking havoc in the cells, the enzymes and their target molecules come into contact with each other and enzymatic reactions take

This enzymatic reaction in mustard can also be perceived in other plant-based foods such as various types of cabbage, horseradish, wasabi, and radish.

The pungency from mustard is due to various glucosinolates: sinalbin from white mustard (*Sinapis alba*, "true mustard") and sinigrin from black mustard seeds (*Brassica nigra*, which actually belongs to the cabbage family). The pungent compound released from sinalbin is rather unstable and the pungency developed from white mustard seeds fades away rather rapidly. After six minutes in pH-neutral conditions only 50% of it is left. However, if the pH is decreased from a neutral 7 to 3.5, for example by the addition of vinegar, then the half-life of the pungent compound will increase to more than 5 hours, so the pungent compounds from sinalbin are somewhat more stable in acidic conditions. Thus, adding vinegar in the beginning of the reaction would significantly help if pungency is desired in the freshly made mustard. The pungent compound released from sinigrin is much more stable and thus black mustard seeds are a better choice for mustards

Myrosinase enzyme liberating pungent compounds in mustard

^ When cells in mustard seeds are crushed, an enzymatic reaction occurs where myrosinase enzyme (blue) cleave glucosinolate molecules (yellow) thereby liberating the sulfur-containing side group as volatile compounds giving a sharp, burning sensation.

that are to stay hot for more than a day. Usually commercial mustard powders are made from a mixture of white and black mustard seeds.

To summarize from mustard and enzyme theory:

1. The pungency of mustard develops via an enzymatic reaction.
2. The reaction has certain minimum requirements, an active enzyme and conditions where it can meet the target molecule, among others. This can occur when cells are broken in the presence of water.
3. Reaction conditions, especially temperature and pH, have significant impacts on the enzymatic reaction and the stability of pungent molecules.
4. It is possible to inhibit enzymatic reactions by adding a suitable inhibitor that fits into the active site of the enzyme.

With these facts in mind, we started to build our experimental setup to learn more about how the theory goes together with the practicalities of making mustard. Since we were tasting very pungent products, we limited the number of samples to three different mustards, as we wanted to avoid scaring the workshop participants away from future workshops. We had to limit ourselves and picked the two questions we thought were most interesting to study:

- How does the reaction time affect the pungency?
- How does destroying the enzyme affect the pungency and overall flavor?

The three versions were based on Tatu's basic recipe.

According to theory, the myrosinase enzyme should be deactivated by the high temperature in Version 1, resulting in a mustard with virtually no pungency. The two next versions should give different degrees of pungency, of which Version 3 should be the most pungent as the enzymes would have had more time to react and produce pungent compounds.

The ten workshop participants were, in a blind tasting, served the mustards and asked to pick the most pungent one, the least pungent one, and which one they preferred most. In addition, they were asked to verbally describe the experiences and flavors given by the different mustards.

Mustard

Ingredients per portion

150 g mustard powder (containing white and black mustard seeds)
150 g water
50 g red wine vinegar
150 g sugar
10 g salt

Procedure according to standard recipe

Heat water and vinegar together. Add sugar and salt and let it dissolve completely. Let the mixture cool down to below 70°C. Mix in the mustard powder and let the mixture rest for 5–15 minutes. Heat to a boil and transfer to glass jar.

Experiment

For our experiments the heating process was modified in the three parallels. The liquids were mixed and heated to a boil, whereby sugar and salt were dissolved into the liquid mixture. Mustard powder was then added to this mixture in three different ways:

Version 1: Mustard powder was added to the hot liquid mixture at 80°C and kept hot under cover for 15 minutes.

Version 2: The liquid mixture was cooled to 30°C, mustard powder was added and kept at this temperature under cover for 5 minutes. The enzymatic reaction was then stopped by quickly heating the mixture to a boil.

Version 3: The liquid mixture was cooled to 30°C, mustard powder was added and kept at this temperature under cover for 15 minutes. The enzymatic reaction was then stopped by quickly heating the mixture to a boil.

The panel was quite agreed in selecting the most and the least pungent mustard. Seven out of ten panel members picked Version 3, the mustard with 15 minutes incubation time with the enzyme, to be the most pungent. A few considered Version 2, the mustard with 5 minutes incubation time with the enzyme, as the most pungent, so these two mustards were perhaps not radically different in terms of pungency. The panel was unanimous in selecting Version 1, the one heated from the very beginning, to be the least pungent of all. This was verbally described as "least mustardy" with descriptions such as "beany," "grassy," "bitter," and "not tasting of mustard." They preferred the two first versions, described as "mustardy," "balanced," and with "good pungency/strength."

What did we learn from this? At least, it became obvious that pungency in mustard can be significantly altered by carefully selecting the reaction conditions, time and temperature. You can prevent the development of pungency completely by heating, thus deactivating the myrosinase enzyme before it gets the opportunity to react. You can also fine tune the level of pungency by controlling the reaction time. In our case both the 5-minute and 15-minute versions were quite pungent. For a milder version, a significantly shorter time, say 30 seconds or 1 minute, might be quite sufficient before heating to a boil. Enzymes are undoubtedly sensitive, but quick if they first get the opportunity to react.

7

The Potatoes that Refused to Become Tender

Red bacalao hotpot is a peculiar dish in several manners. Firstly, it is both Norwegian and Spanish/Portuguese at the same time. In Norway it dates back to the 19th Century when cod traders from Europe introduced it. The dry-salted codfish, *klippfisk*, for the dish is definitely Norwegian, produced in western parts of Norway. In the high North, dry codfish, stockfish, can be dried by hanging them from wooden racks without the use of salt, but further south the humid climate makes it difficult to dry large fish without salting it heavily. The method for salting and drying the fish was imported by the Dutch trader Jappe Ippes around 1690. He taught Norwegians who lived along the Atlantic coast how to make the cod last months, in order to be transported by ship to southern Europe, South America, and the Caribbean Islands, much the same way as the Basques and Canadians of Newfoundland did. The other important ingredients of the red bacalao hotpot based on this saltfish are Mediterranean as well: tomatoes, pimientos (canned bell peppers), cayenne pepper, olives, and olive oil. This delicious dish would never have seen the light of day if we didn't have an extensive cross-cultural communication!

Red bacalao hotpot

There exist many versions of this dish. Some families have their own recipes that are passed down through the generations. The recipe below is a fairly common version, enough for at least four diners.

Ingredients

900 g klippfisk (salt cod)
750 g potatoes, peeled and cut in 1/2 cm thick slices
2 large onions, sliced
3 red bell peppers or a can of tinned pimentos, diced in 1 × 1 cm pieces
2 dl olive oil
3 Tbsp. seedless black olives
4 cans (4 × 400 g) tinned tomatoes, preferably coarsely cut
2 cloves garlic, finely chopped
1 red chili (medium hot variety) finely chopped without seeds, or cayenne pepper to taste
2 bay leaves
2 Tbsp. parsley, chopped

Procedure

1–2 days before serving: After rinsing off excess salt under cold running water, cut the fish in 5 × 5 cm pieces and leave in cold water for at least 24 hours. A rule of thumb is 24 hours per cm thickness of the loin, changing the water a couple of times a day.

On day of serving: Sauté garlic, chili and bay leaves in some oil in a large pan. Starting with the fish, place fish, potatoes, onions, and peppers on top of each other in layers going several rounds of layers. Pour the tomatoes over this. Heat to a boil and let simmer gently until the potatoes are tender. Various recipes specify cooking times from 45 minutes to 1 hour 15 minutes. Shake the pan once in a while, avoid stirring. Garnish with good olive oil and parsley, serve with good bread.

Bacalao is a fairly common dish in northwestern Norway, and cooking it at home usually comes out well every time. As it is easy to prepare in large quantities at a time, it is also a common food when large groups of people are to be served. Bacalao is delicious and the cook succeeds with it almost every time. The most common failures are stirring too much so that the pieces of fish crumble, or to overcook it so that the potatoes turn mushy. The reason for the latter can be too long cooking time or a potato variety that is too mealy. Otherwise this dish is more or less foolproof.

Therefore, we were more than interested when hearing a story about a "bacalao failure" that was quite the opposite: a group of non-professional, but experienced, cooks had prepared a huge pan of bacalao to be sold at a school brass band festival. They started preparing the ingredients several hours beforehand, heating the stew for it to be ready when the musicians, parents and other spectators were to arrive. However, this time, the potato slices refused to become tender, even though they had been simmering for more than an hour! After all, cooking to the texture of *al dente*, slightly crunchy, is seldom what you aim at when boiling potatoes. What had happened?

About potatoes turning soft during cooking. The structure of a cooked potato is usually somewhere between mealy and firm (waxy). This is a result of both potato variety and cooking procedure. The mealy varieties, often deep fried or baked in the oven, commonly give mealy and even fluffy potatoes with a dry mouthfeel, whereas the waxy varieties, cooked in water, should feel moist and almost a bit juicy. However, most importantly, when we want to eat cooked potatoes we expect the interior to be soft, regardless of whether it is cooked in water or milk, fried in oil or baked in the air of the oven. A firm "crunchiness" usually indicates that the potato is raw, undercooked. Keep on cooking!

Potatoes soften during cooking when the starch granules inside the plant cells during heating take up water and swell—the starch gelatinizes. The cell walls also start to soften and cells originally glued together start loosening, or even detach, from each other. A major reason for the crunchy structure of the potato is this glue, pectin, located between the cells. If the pectin is strong and active, the cells are firmly attached to each other, and when you bite into it, it feels

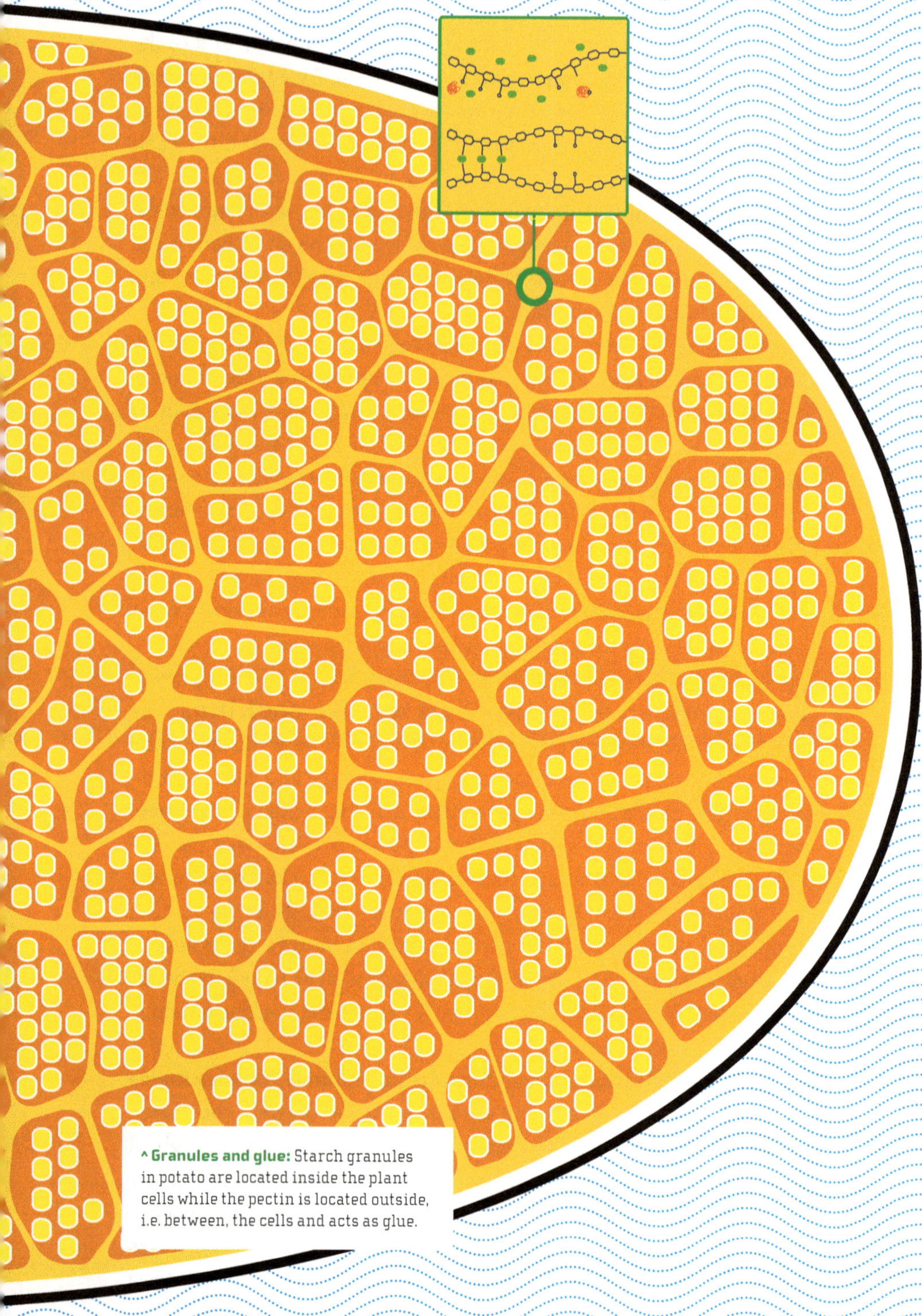

^ **Granules and glue:** Starch granules in potato are located inside the plant cells while the pectin is located outside, i.e. between, the cells and acts as glue.

firm and even crunchy. A sensation that suits well when taking a bite of apple or carrot, but never for potato. Luckily, pectin is not soluble in cold water. If that was the case, potatoes and other fruits and vegetables would have started dissolving upon the first contact with water. Most likely, they wouldn't grow to become vegetables and fruits in the first place. Pectin is, however, soluble in hot water, and when vegetables and fruits are heated the pectin is partially broken down and dissolved. Consequently, the cells loosen from each other and the potato becomes softer, and eventually breaks apart into pieces. When you bite into it, you break off groups of more or less intact potato cells. Taken to the extreme, all cells detach from each other and you have a potato soup. Even when baked in the oven, there is enough water in the potato itself to dissolve the pectin, but due to absorption of free water by starch granules and some water evaporation, oven-baked potatoes dry out along the way. The other extreme of a cooked potato is that the pectin glue remains strong and the cells remain tightly bound together. Thus, the teeth need to do the hard work and rip apart, or break up, the potato cells: the potato feels crunchy.

The mystery of the persistent potatoes. Firm potatoes are recommended for bacalao because potatoes that are too soft would give a cloudy, mushy stew which is outside the style of the dish. However, even firm potatoes should be soft enough not to be felt as crunchy. At the brass band festival, the pectin apparently did not dissolve as planned. A food chemist would start looking for possible explanations to such a mystery. We envisaged at least three hypotheses to explain why potatoes in a bacalao hotpot would remain firm even after prolonged cooking.

Firstly, it is a well-known fact that softening of vegetables during cooking takes longer under acidic conditions. This is, however, not due to pectin but an important component responsible for rigidity of cell walls, namely hemicellulose. Since bacalao contains tomatoes, it is likely that the bacalao cooking liquid is acidic and might have retarded softening of the potatoes. But the potato problem in question first appeared at the festival using a fairly standard recipe, among experienced bacalao home cooks who had not experienced such problems previously. If acidity from tomatoes was the culprit, the cooks should have experienced this when cooking at home as well, and this should be a common problem cooking bacalao. It is not, and no recipes to our knowledge mention this as a potential problem.

The second possibility that occurred to us was the tendency of pectin to turn into a non-soluble form in the presence of certain metal ions from naturally occurring minerals, such as calcium or magnesium. Pectin, which, like starch, belong to the carbohydrates, is a long polymeric molecule where different organic acids and sugars form long chains. Under certain circumstances, the organic acids of the pectin chain can interact with calcium ions present in water. One calcium ion can interact with two acid groups to form bridges between the long pectin molecules. Such cross-links cause the pectin to form stronger networks making it less water soluble. When enough such bridges are formed, the pectin becomes practically insoluble, even in hot water. Vegetables cooked in calcium-rich media such as milk, cream, or hard water take longer to cook. They may even remain hard after very long boiling. Could it be that the water at the site was particularly hard, rich in minerals? We did not see this as a plausible reason because the water in the area is generally quite soft and some of the cooks indeed lived in the area and had not experienced this problem at home. But since we did not analyze the water, the hypothesis could not be completely ruled out.

The third hypothesis is closely linked to the previous one, but takes the cooking temperature into account. In plant cells there is a certain enzyme called pectin methylesterase, PME, which can cut off certain parts of the pectin molecule making it easier to associate with calcium. Pectin molecules are cross-linked to form networks, much like the hardening that occurs in glue (see illustration). The extensive pectin network becomes poorly soluble even in hot water, and the cells stay together resisting the teeth of the diners.

The optimal conditions for PME action are a certain temperature, ideally between 60 and 70°C, and the presence of calcium or certain other metal ions. Since calcium is present in all living cells, it is sufficient to keep the temperature between 60 and 70°C for a certain amount of time for the pectin to be hardened. Actually, PME works 100 times faster at 65°C compared to room temperature, so keeping potatoes at this temperature for 15–20 minutes is sufficient for the enzymes to do their job. Perhaps this was what happened with the persevering bacalao potatoes? Maybe the large pan took so long to heat that the potato slices were subjected to temperature and time for PME to crosslink enough pectin to harden the potatoes?

The experiment. We had three possible hypotheses. The first one was quickly falsified, so we didn't find any reason to follow this line of inquiry into an experiment. The second, minerals such as calcium, was difficult to test because we did not have access to the water used in the mentioned case. However, follow us to the test of the third hypothesis and we will try to convince you that, in this specific case of bacalao, the second one can also be disproved.

We conducted the experiment as part of a food workshop at a Norwegian culinary school with culinary students, teachers, and other guests as participants. The teachers and students cooked three pans of bacalao based on the same recipe. Only one factor varied: the temperature and time in which the potatoes were cooked. In the first parallel, we cooked the bacalao the traditional way, exactly according to the recipe. In the second, we chose cooking time and temperature to deactivate PME. Since enzymes are a subgroup of proteins, and behave as such, they denature (and lose their ability to function) if subjected to higher temperatures, commonly above 70°C. The potato slices were therefore blanched: they were plunged in boiling water and cooked for four minutes, whereby they were transferred to ice water to stop the cooking process. The bacalao was then cooked according to the recipe using these blanched potatoes. In the third parallel, we chose conditions that should promote PME activity. The sliced potatoes were transferred to a pan of 65°C water and the temperature kept as stable as possible for 15 minutes, whereby they were used to cook bacalao according to the same recipe and procedure as the first two parallels.

The verdict. The three ready-cooked pans of bacalao were blind tasted by the workshop participants and evaluated according to several pre-defined criteria. We asked them especially to point out the two samples where the potatoes were hardest and softest, respectively. The taste panel of the day reported that the potatoes in the third parallel had the firmest potatoes. That was the one where we had attempted to promote PME action and reinforce pectin to achieve firm potatoes. This parallel was perceived as having the least crumbly and

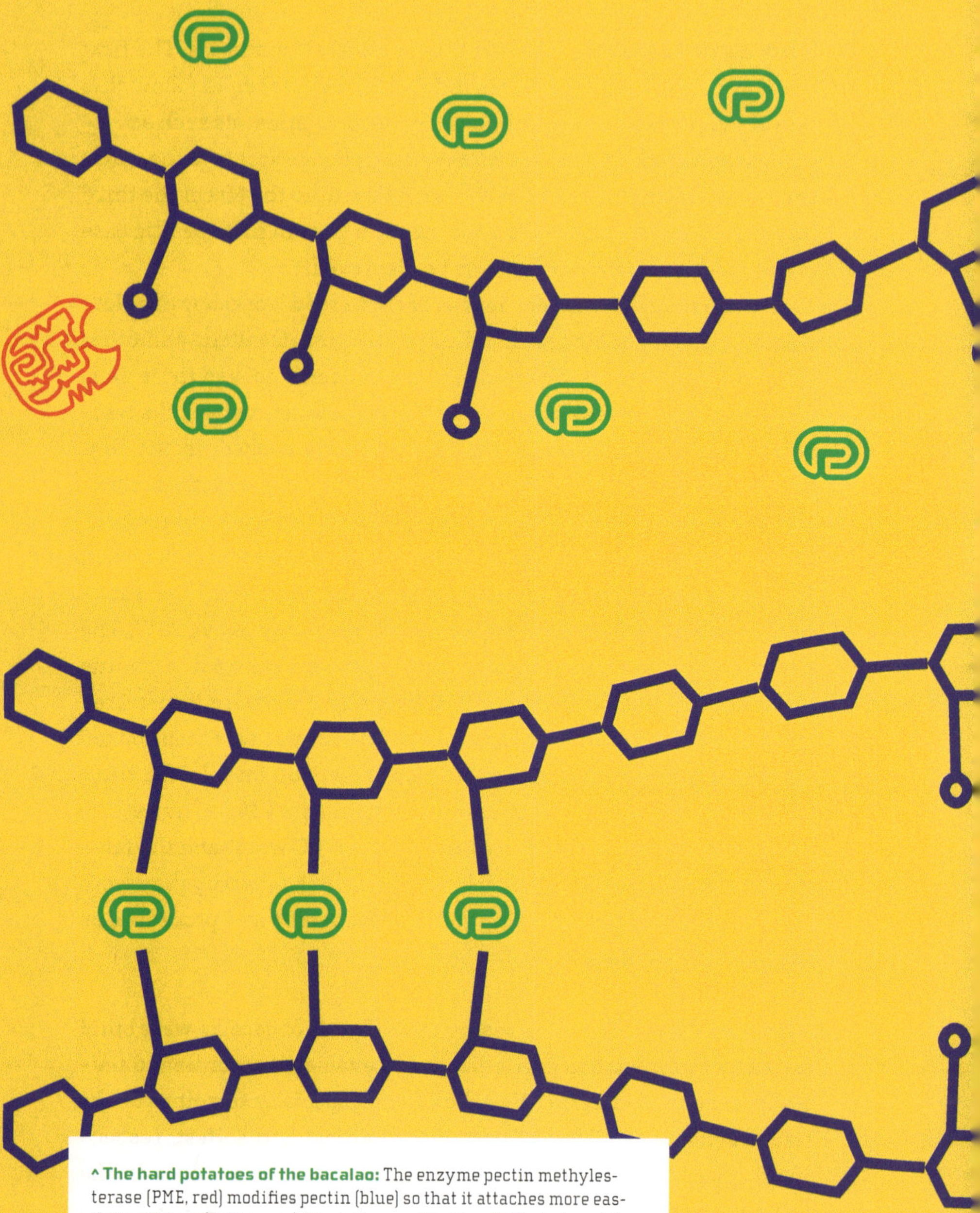

^ **The hard potatoes of the bacalao:** The enzyme pectin methylesterase (PME, red) modifies pectin (blue) so that it attaches more easily to calcium (Ca^{2+}, green). The latter can bind two PME-chains at the same time, thereby forming a strong, hardened network. These potatoes will not become tender even after extended time in boiling water.

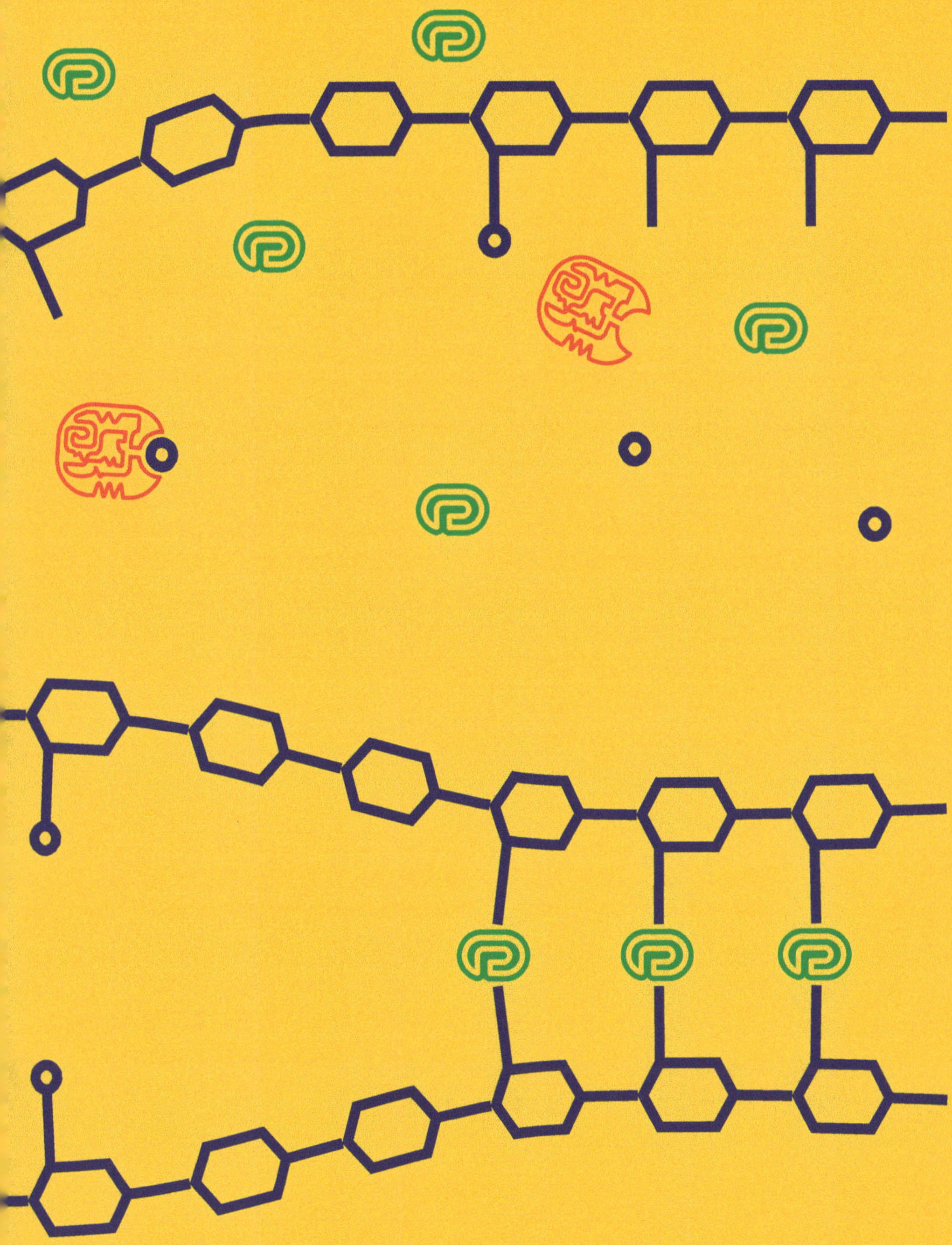

Usually we experience that cooking small pieces of food takes shorter than large pieces. However, occasionally the opposite may happen. A friend once wanted to speed up cooking potatoes by grating them into the smallest possible pieces. Surprisingly, the grated potatoes never became tender. It would have actually been quicker to cook them as whole potatoes. PME is most likely to blame for this, with grating enhancing the effect. In plant cells, PME and calcium are kept separate, located in different parts of the cells. But when grating the potato, cutting thin slices, or maybe even using a blunt knife, cells are broken and the two come in contact and can act even more efficiently to cross-link pectin and harden the potatoes. When boiling potatoes whole, the same happens because cells swell in warm water and some of them burst open. So, if your knife is blunt, you cut thinly, or grate a vegetable, be prepared for PME to firm up your food. If you don't want this to happen, make sure the food meets high temperatures quite quickly. The mystery of potatoes refusing to soften is known in Finland as well. Many Finns are familiar with grandma's advice that new (young) potatoes should be placed in the pot only **after** the water is boiling: "If you put new potatoes into cold water, they will never soften properly!" Fully grown may be put into the pot of cold water before turning the heat on. The reason for this phenomenon is the same as for bacalao: the pectin hardening enzyme PME. New potatoes, unlike stored potatoes, are still in their growing stage, as they are harvested early in the summer during the best growing season. In this stage, the enzyme activity is very high and the PME is also more active than in potatoes that have ripened and been stored during the harvest season. ×

least smooth texture. The second parallel, where we had tried to deactivate the PME enzyme, was evaluated as having the softest, smoothest, and most crumbly potatoes. Indeed, we had got two extremes on the potato texture scale, while the bacalao cooked according to the classic recipe came out in the middle on all descriptive accounts. Our suspicion that PME was indeed the reason for the experience during the festival was considerably strengthened. The result would obviously not stand as a proper research study, but we had a fairly clear indication.

Evaluating which is hardest or softest are rather descriptive criteria, not directly linked to one's personal taste or preference. So, taste may indeed be used to make fairly objective evaluations and people are able to agree upon issues of taste if the task is clear and associated with something descriptive. However, when we asked which version of the bacalaos the participants preferred, the result was, as often, not as clear cut. Taste evidently differs among people, and, interestingly, the traditional way of preparing bacalao was the one to receive the highest number of votes. Who would have guessed. Science and analyses can explain, but often art or craftsmanship will produce what is most enjoyable.

8

Gravlax and a Pinch of Salt Chemistry

Microorganisms—yeasts, molds and various bacteria—are everywhere in our environment. They are in the ground, in water and in the air. On the floor, on the table, and at the doorstep. On the skin, hands and under your nails. Microorganisms are on the surface of food, be it meat, fish, or plant product. Even the surface of each flour particle is populated by microorganisms. Fortunately, the vast majority of them are not harmful to humans, and many are even beneficial for us. However, they are ready to pop our food into their own mouths, insofar as microorganisms can be said to have a mouth. No wonder that fermentation, modification of foodstuffs by means of microorganisms, is so common in all parts of the world. In practice, all types of raw materials have been, or are being, fermented: fruits, vegetables, cereals, meat, and fish. Most likely, fermentation has been in use as long as humans have existed.

Although common in the ancient Roman kitchen, the important Garum, a strong and smelly fermented fish sauce, disappeared from European kitchens for over a thousand years, but has now again emerged in modern kitchens similar to the likewise fermented Vietnamese fish sauce. From the late mediaeval times, fermentation was a

as well. There is a great number of "original" gravlax recipes in the food literature and cookbooks, and the variations seem to have some regional character. Finns often marinate their salmon with only sea salt and flavor it with dill. If there is sugar in a Finnish gravlax recipe, the amount is moderate. Swedes and Norwegians always add sugar, from half to equal amounts compared with salt. "Why can't they do it the right way, as the Finns do?" Anu the Finn asks. Maybe we should simply let the Danes, Finns, Icelanders, Norwegians, and Swedes continue their everlasting friendly debate over whose gravlax is the best. Preferences are personal and the familiar one is often considered the best. However, what we all can share is our interest on the phenomena taking place during, and the possible explanations behind, the "graving" process. One procedure that seems to be common to most recipes regardless of origin is the procedure that follows immediately after seasoning: the boneless fillets, skin on, are sprinkled on the meat side with the other ingredients, whereby they are placed against each other meat sides facing together and skin sides out. They are tightly wrapped and left to mature in the fridge. Obviously, one fillet must lie underneath, the other on top. So, the question arises: should the fillets be flipped to switch places during the process, or not? Some recipes tell you to flip the fish around halfway during the process while others do not mention this. Tatu the chef had reflected upon this, and one day he asked the very opportune question: "Does salt climb?" What an excellent question! In effect: does it make any difference if the fillets are flipped over? Do we get two different types of gravlax in one if we don't flip?

We decided to test the flipping issue. Tatu bought a good fresh salmon from the local market hall and cut out two identical fillets of each 620 g. Both fillets were sprinkled with 30 g sea salt and 30 g sugar on the meat side and were put on top of each other, meat side in and salt/sugar in between. The following 50 hours were used as each found most suitable, before we met for a workshop to evaluate the result.

A pinch of salt chemistry. Table salt, sodium chloride, is not a molecule. It is a salt, among a countless number of different possible salts. Sodium chloride is composed of two different types of ions with opposite charges: sodium ions with a positive charge, chloride ions with a negative charge. In dry environment, these two attract each other, a bit like the opposite poles of two magnets. These forces give a well-organized three-dimensional structure with every other ion being sodium and chloride. The result is a beautiful octahedral structure, which also is reflected in the macroscopic structure of salt crystals visible to the naked eye.

The electrostatic attraction between the two different ions is very strong. However, water molecules have the ability to penetrate between the sodium and chloride ions and jiggle them apart, one by one. This way, salt dissolves in water, each ion surrounded by one or more layers of water molecules, to start their individual journey in the watery world of food.

Salt dissolves in water

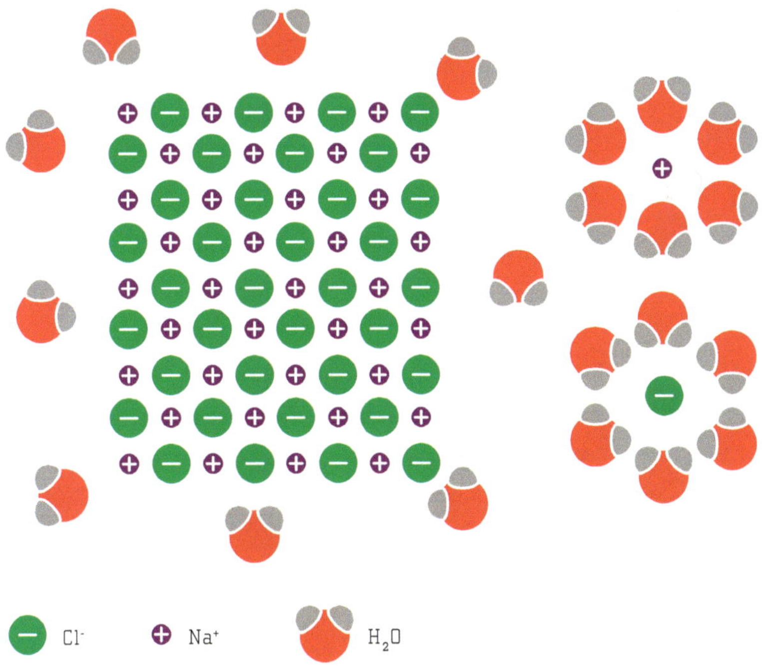

^ **Salt dissolves:** Each sodium and chloride ion requires six water molecules to tear itself away from the salt crystal and flow freely around in the water. The salt is completely dissolved when all ions float around individually.

Salmon fillets with salt

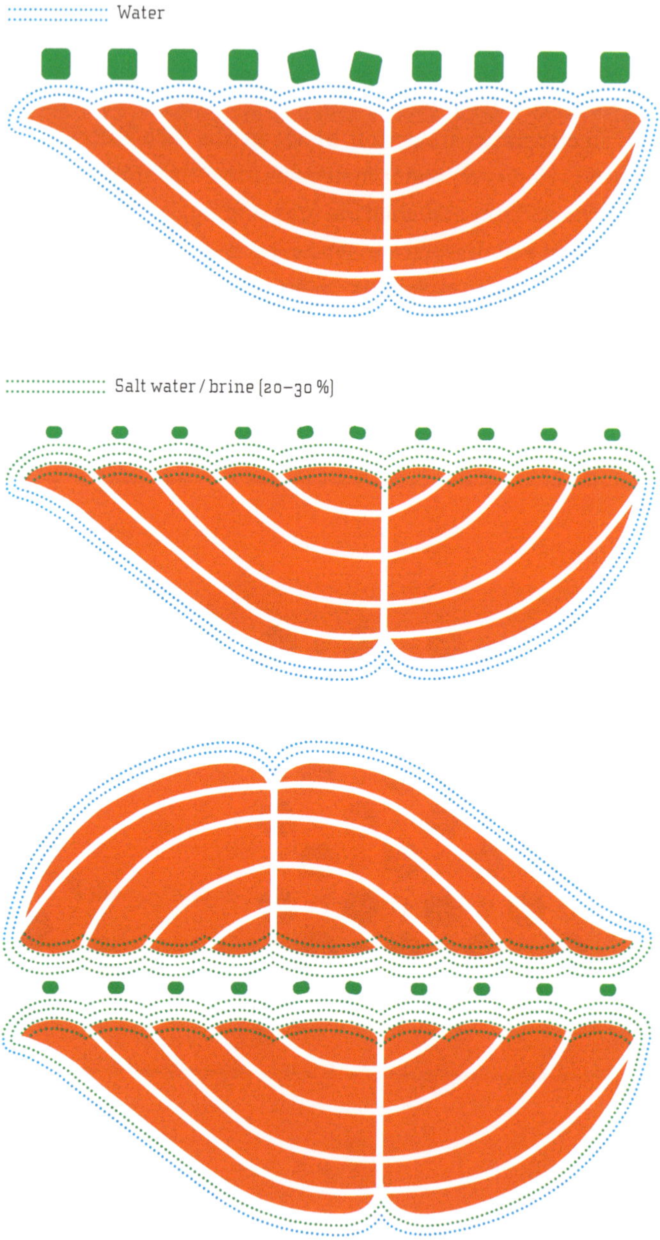

^ Diffusion of salt: When salt dissolves at the surface of the fillet, the salt concentration is much higher at the surface compared with the inner parts of the meat. In time, dissolved salt ions will spread—diffuse—into the meat through tiny channels.

Salt dissolved in water moves around randomly, entirely based on collisions in the submicroscopic world of ions and molecules. Over time this will result in salt from areas of higher concentration to distribute more evenly throughout the water—diffusion.

When salt is dissolved into the thin water layer that covers the food, the saline concentration on the surface is much higher than inside the food and salt ions diffuse slowly into the regions with initially lower salt concentrations wherever there are channels and gaps.

However, salt is not the only thing that moves around in food. Water inside and in between muscle cells also diffuses and moves as result of these random movements. Only when the food is completely frozen, is the movement halted or severely slowed down. In sum, salt moves inwards toward areas of low salt concentration while water moves outwards toward areas of lower water concentration (due to the high concentration of salt on the surface). In a typical gravlax, the fish loses water and approximately 4% of its weight. Salt also affects the fish structure. The salt ions change the electric charge of the muscle fibers, forcing them apart from each other, which again results in a fish that feels moister and more tender in our mouth. A more tender muscle that has dried out sounds like a paradox, but it is indeed the case.

Back with Tatu's gravlax, our theoretical pondering indicated to us that it was unlikely that the upper fillet would mature/marinate differently than its lower counterpart as long as these are purely chemical phenomena. Gravity is no match against the forces of chemistry; the diffusion rate is the same whether something diffuses upwards or downwards in a solution. However, the saline water released from the fish tissue will flow downwards, and thus it is plausible that the fillet at the bottom experiences a more intensive brine bath than the one on top. But would it be possible to taste any difference? Tatu rinsed, dried and weighed the two salmon fillets, sliced them and coded them for blind tasting. Each workshop participant received a slice from each fillet, cut from the same place in the two fillets.

The taste panel members were asked to evaluate which of the two was saltier, which was juicier, and which was more tender. They were also asked to choose which of the two they would prefer and why. There was practically no difference in weight loss between the two fillets, both had lost approximately 4% of their original weight.

In the blind tasting, they were also evaluated as equally salty, that is, the panel was divided as close to the middle as possible as to which was most salty. So, does it make any difference if the fillet lies on top or underneath? Indeed, it does! The majority of the votes, nine out of 13, described the lowermost fillet as both juicier and more tender than the topmost. In their descriptive accounts, the panelists described the uppermost fillet as firmer, with a denser structure and more evenly distributed salt. The lowermost was described as sweeter, having stronger fish flavor, which also released more liquid. When asked which they preferred, the uppermost was the winner with nine votes. Interestingly, some preferred the uppermost because it was less salty while others preferred the lowermost because it was more salty. For some, the preference comes down to their personal liking in terms of saltiness.

If we are to trust these results, there is a difference between the two treatments, and the upper fillet seems to mature slightly slower than the lower one. That is, a more rapid "graving" process in the lowermost fillet, which also experienced a steady slow stream of salty juices from above. We would guess, or hypothesize, that turning the salmon fillets halfway through the process may give a more even results in the two fillets. And if you do not flip them, you will make two types of gravlax with one single procedure. And, at the same time, you have ensured something to talk about with your guests when you ask them if they would prefer having their slices from the upper fillet or lower fillet.

9

The Art of Heating

Cooking very often implies heating, that is, increasing the temperature of the food. Temperature as phenomenon basically comes down to one thing: movement. In gases and liquids, such as the hot air in an oven or water in a pan on the stovetop, the temperature directly reflects how fast the molecules are *moving*. In solids, where the molecules are kept firmly in their places not moving around (in fact, this is what makes an object or substance a solid), the temperature reflects how intensely the atoms or molecules *vibrate*. So, whenever we measure temperature in a material, we usually measure movement at the molecular level. Cook with heat therefore implies transferring energy to the food. The source of this energy can be the surface of a skillet, the water or gravy in a pan, the air in an oven, the radiating elements of a microwave oven, the coals of a barbecue grill or the rocks in a cooking pit.

The choice of energy source will contribute to how the food ends up: a pork chop can be simmered, shallow fried, oven baked, or grilled. If you ask a physicist, they would probably say that it is this transfer of energy that is called "heat." Such thermal energy spontaneously spreads from a place with high temperature to a place with lower temperature. According to physics, this energy transport can occur in two different manners: *conduction* and *radiation*. Additionally, energy transfer in cooking can benefit from a couple of other concepts: *convection* and *condensation*. These four concepts can be quite useful when we talk and write about food and cooking, so let's try to put some meaning into them.

Conduction—when hot and cold surfaces touch. One of the most striking experiences of conduction that probably every human being at some point experiences is when you burn your hand as you accidentally touch the ovenproof dish in your oven, or place your hand on the hot skillet. Conduction occurs when two objects are in direct contact with each other and one has higher temperature than the other. The molecules in the skillet, which is of solid metal, are vibrating intensely. When these come in contact with the surface of some food, or your hand, they collide with high intensity into the molecules on the surface of the food (hand), thus transferring some of their energy by setting the molecules of the food into motion. If the food remains in the hot skillet, a domino effect occurs from the outer parts inwards. This domino effect, a gradual conduction between outer and inner parts of the food, is one reason for the food eventually becoming hot all the way through. Substances that are efficient heat conductors are consequently efficient for heating food. A metal, YES. Styrofoam, NO.

Some metals are better conductors than others. If you have had the pleasure of stirring a cup of hot tea or coffee with a silver spoon, you might have noticed that the spoon feels hotter than your everyday steel spoon. This is because silver conducts heat much more efficiently, by this domino effect conduction, through the length of the spoon to your fingers. On the other hand, when one of our prehistoric forefathers made a primitive skillet by placing a flat rock on top of a burning fire, it was rather inefficient compared with the metal pans used in the later metal ages. S/he might be able to heat the rock to fry one side of the meat, but the rock is not very efficient in continuously transferring the heat from the fire to the food, so s/he would have to move the meat around to new spots every time the meat was turned.

Liquids and gases are comparably poor conductors of heat, one reason being that the distance between the molecules is larger than in solids. So, the collisions between the cooking medium (gas or liquid) and the food are necessarily less frequent compared with cooking on a solid surface such as a metal. It takes longer to heat food in an oven,

Relative conductivity compared with water for selected materials

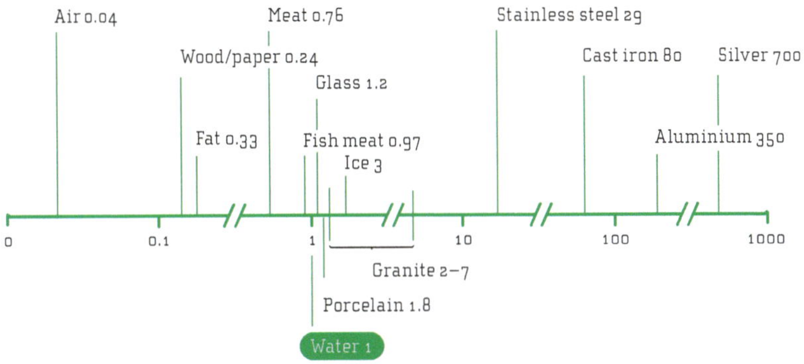

^ **From air to silver:** The table shows relative conductivity for various materials relative to water. Metals are good conductors, but even among these the differences are large.

where the heat transfer medium is air, compared with having the food in direct contact with the hot metal surface of a skillet. Since water has much higher density than air, water is a much more efficient heat transfer medium; the molecules are more densely packed and collide more often with your food. So, cooking potatoes in water is usually quicker than baking them in the oven, even though the temperature in the oven might be much higher than in the water, boiling at 100°C.

A barbecue radiates. Feel the sun warming an otherwise chilly, spring day, or feel the warmth sitting beside your wood-fired oven. In both cases, you experience thermal radiation warm your cold toes. Heat through radiation is different from conduction in that there needn't be any physical contact between what is hotter and colder. In radiation, energy moves via electromagnetic waves, a form of light. If we leave out microwaving for a while, it is basically infrared radiation we use when we use radiation for cooking. The radiation is absorbed at the surface of the food, where molecules are set into motion, and the food is heated. The energy source (the coals in a barbecue, the hot walls or broiling elements of an oven) throws packages of energy around in all directions. When a molecule at the surface of the food is hit by one of these, it is energized and starts vibrating or moving around faster. An example of a food that is cooked purely by infrared radiation is the Finnish way of cooking salmon called "loimulohi," flame-roasted salmon. You fasten a salmon fillet to a plank by driving wooden plugs through it and then place the plank upright by the side of the fire. If you don't place it too close to the fire, the only heat reaching the fish is the radiation. The hot air from the fire move upwards and will not reach the food (see convection below).

Microwave cooking also relies on radiation but in a somewhat different fashion than infrared radiation from hot bodies. While infrared radiation affects basically all types of molecules and materials, microwaves are more selective. They are particularly efficient in setting water molecules, and, to a certain degree, fat molecules, into rotating movement. Consequently, almost all of the infrared radiation is absorbed at the surface layers of the food, while microwaves may travel several centimeters into the food before being absorbed. Thus, food that is microwaved is heated just as much from the inside out as from the outside in. When a molecule absorbs an energy package in the microwave frequency range, it is set into intense motion. This

In his book, *What Einstein Told His Cook*, the American chemistry professor asks: On a really hot day, would it be possible to fry an egg on the bonnet of a car? The answer would be "probably not." The reason is the fact that as soon the egg lands on the bonnet, the surface will be cooled while the egg becomes hot. But since there is no other heat source than the sun radiating energy onto the bonnet, you end up with the egg partially fried. However, if you opened the bonnet and cracked the egg straight onto the engine block, chances would be considerably better for frying the egg all the way through, especially if the engine was running. Having the engine running would be similar to cooking on the stovetop, the combustion process constantly feeding energy to the metal. The amount of metal in the engine block is also much higher than in the bonnet, more energy would have been stored in the material, which would in turn be transferred to the egg to cook it. ×

results in small areas of very high temperature. Therefore, you would want to wait for a while for the energy to spread from the hot spots to the cooler regions in your food via conduction.

Convection—boiling and baking. If you burn yourself holding your hand over a candle you experience energy transfer via convection. The phenomenon occurs when a liquid or gas of higher temperature moves to a place of initially lower temperature. In the case of the candle, warm gases flow upwards from the flame to reach your hand, after which your hand becomes heated by conduction when the hot gas comes in physical contact with your skin. Since convection is all about "hot molecules moving to new places," it can only occur in media that are able to flow: liquids such as water or oil, or gases such as air or steam. Since these "hot molecules" have high energy, they move about or rotate in high speed, and when they collide with "colder" and slower molecules, some of their energy is transferred via conduction. Convection processes are common phenomena both in nature and in everyday life at home. While heating water for your pasta, you may see flows and currents in the water even though the pan stands perfectly still on the hotplate: hot water from the bottom rises and pushes colder water to the side, and so the circulation goes. Warm air rises while cold air sinks both as a meteorological phenomenon as well in your own living room.

The difference between conduction and convection can be illustrated using a dance floor as metaphor. Envisage a dance floor where some (colder) couple dance slowly cheek to cheek while other (hotter) couples dance a more intense swing, as shown in the illustration. The red steps illustrate fast swing, "hot molecules." The blue pairs dance slowly and are "colder."

During conduction, the cold couples surrounding the hot couple are affected, inspired, as they come in contact with the latter. After a while, the cold couples become hotter, while the originally hot couple has shared some of its energy and become somewhat cooler. This illustrates conduction in both solids (all maintain their original positions, but can still rotate on the spot and bump into each other) and in liquids and gases (the ones surrounding can move around when pushed by the "hot" couple).

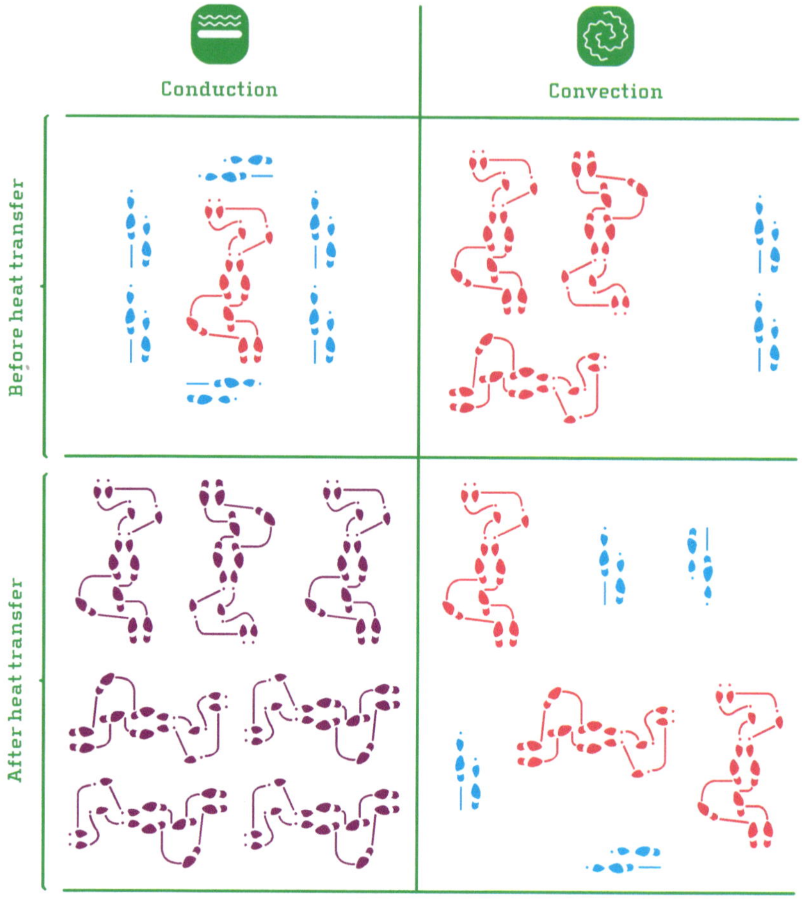

^ Conduction and convection illustrated by hot and cold couples on the dance floor. Red steps represent hot molecules, blue represents cold and purple represents temperatures between hot and cold.

Convection is illustrated by a room with one cold end, inhabited by couples dancing slowly, "cold molecules," and the other end dominated by more energetic dance. Very similar to a hot oven standing in an otherwise cold room. Since molecules by nature move randomly and are reflected when they bounce against each other or a surface, they will in time distribute evenly across the room. We move toward an even temperature in the whole room for two reasons: 1) the various couples will over time distribute evenly throughout the room. 2) The "hot" couples will eventually bounce into the "cold" ones, sharing their energy (this is consequently conduction).

The most striking practical example of convection when cooking food is when you turn on the fan in your oven; indeed, ovens with such fans are called "convection ovens." Many ready-made foods have instructions that you should heat at different temperatures depending on whether you use the fan or not, 200°C in ordinary ovens or 180°C with the fan running. This reflects the fact that the fan sets the air into motion to increase heating through convection. Even without the fan, convection also occurs via hot air moving out from the hot surfaces, pushing aside less hot air.

When barbecuing/grilling, at least over a campfire, you have the choice of holding your hot dog on the side of the coals or above them. Everyone having roasted hot dogs by a campfire has experienced that holding them above the coals is more efficient. Hot air moves upwards from the fire, and the hot dogs experience convection in addition to radiation. This is a more intense way of heating, which often results in a hot dog with a burnt surface while still cold in the middle. Apparently, sometimes convection can be a bit too effective way to transfer energy. However, if you hold the hot dog by the side of the fire or coals, which is also recommended as you avoid transfer of soot particles from a burning fire, you would be heating by radiation from the flames or embers, just like cooking Finnish loimulohi.

Condensation—steam in a closed pan. Rather than plunging your potatoes into hot water, you may place them on a grate with just a little water in the bottom of the pan. You place a glass lid on top, turn on the power. Your potatoes are ready almost just as quickly as when boiling them in water, without many visible signs of anything happening in or around the potatoes, possible except for some water dripping off the potatoes. How come?

Let's return to the burnt hand because that experience is quite a powerful association for many of us. If you place your hand in front of the spout of a kettle with boiling water, you may be badly burnt. You will not only experience convection and conduction from 100°C air and steam reaching your hand transferring energy by contact (conduction), but in addition the steam, water in form of gas, will condense back to liquid water on your skin. It requires a large amount of energy to make water evaporate; indeed, it takes seven times more energy to evaporate water (already holding 100°C), as it takes heating the same amount of water from room temperature to 100°C. All the energy that went into evaporating the water is returned to your skin when the steam reaches your hand and condenses to become small drops of liquid water. This comes in addition to the hot water transferring its heat energy by ordinary conduction from temperature difference, like if you had stuck your finger into hot water. This is also the reason that you are more severely burnt if you come in contact with steam at 100°C compared with dry air of the same temperature. After all, you can stick your arm into an oven holding 200°C without getting burnt, but holding your hand just above a boiling pan of water will cause severe burns at "only" 100°C. However, what you actually see coming out of the kettle spout, or of a boiling pan, is not steam. Water vapor is a colorless and invisible gas. What you see is liquid water, steam that has already condensed into miniscule drops on its way out in the kitchen. If you look closely at the tip of the spout, or just above the boiling water, you would see (apparently) nothing. *That* area is where the water vapor is, and that is the place where the energy transfer to other bodies in the most intense.

When you steam food, you make use of this evaporation–condensation energy. Steam moves by convection from the boiling water below your food, hitting the surface of the food to release both condensation energy as well as some heat energy through direct contact, conduction. Consequently, steaming is quite efficient and competes well with boiling it in water even though steam is much less dense than liquid water. But don't lift off the lid to have a look every second minute because this will release the efficient steam out into the room. A transparent glass lid is a good thing for the curious cook. If the water in your pot is boiling but you don't see white clouds, you can be fairly certain that the pan is full of invisible steam condensing on your food to cook it.

Cooking with heat. It is very seldom that only one form of heat is in action when cooking. Two exceptions mentioned are the Finnish "flame-roasted" loimulohi, and microwaving, where radiation is the only source for heating. But even in these cases, conduction, convection, and condensation occur inside the food: the heat spreads and the temperature evens out throughout the food. Usually, several or all forms of heat are in action at the same time.

How often should you turn a steak in a skillet? This important question arises when pan frying, because there the heat transfer takes place both via conduction and convection: while the heat is conducted from the hot surface of the skillet to the food, energy is transferred by convection from the hot surface of the steak to cooler, surrounding air. The steak also cools via evaporation of water from the steak surface, the opposite of condensation. So at least three different heat transfer mechanisms are involved in the frying of a steak (in practice, there will also be radiation from the skillet surface onto the meat). Traditionally, many of us have learnt to turn the steak once during cooking—when you see juice seeping out on the top. But lately, chefs have started turning it more frequently, as often as every 30 seconds. An explanation for the superiority of this new technique lies in the optimization of heat transfer. If you turn the steak only once, the hot surface of the turned steak will cool down longer and more compared to the frequently turned steak, and the total cooking time is increased compared with the frequently turned steak. When turning frequently, the hot surface cools more evenly, avoiding significant temperature differences between the outer and inner parts. Furthermore, it has been found that the layer 1–3 mm beneath the surface is heated to about 10°C higher if turned only once compared with the frequent-turned steak, risking overcooking of the outer parts. In other words, flipping your steak frequently would both reduce cooking time and promote a more evenly cooked steak. ×

In a coal barbecue, the coals are very hot and the heat transfer occurs via radiation plus some convection by hot, rising air. But where the food is in direct contact with the grates, conduction occurs between the metal and the food. This heating is sufficiently intense to give the barbecued food its appealing grated pattern, also giving a more complex and appealing flavor. If you have a gas barbecue, the radiation is not as intense as in the glowing embers, and you get comparably more convection compared with radiation. The resulting process is somewhere between a coal barbecue and baking in an oven, particularly if you keep the lid on. In an oven, the surfaces are not as hot as the embers of a barbecue, so the radiation from the walls are less intense. However, the air in the oven is hot and kept locked inside the confined space, and, therefore, the most important factor when cooking in an oven, unless you have the grill function turned on, is convection in air.

Although it might seem self-evident, it is worth reflecting upon the fact that heat almost always moves from the outside of the food inwards when we cook, the seemingly only exception being microwaving. If we use high temperature when cooking a large piece of food, the outer parts will be overcooked before the heat reaches the inner parts. Keeping the temperature lower will make the cooking take longer, but the domino effect of heating the food throughout will have more time to occur without the outer parts becoming overheated. This is one of the main principles behind the sous vide technique, where the food is vacuumed and left in a water bath of desired temperature for a long time, allowing the heat to penetrate the food more gently. Chemical reactions are allowed to proceed, giving good flavor and tender meat. We will, however, not go into sous vide cooking here as great sources to information about this are only a few clicks away on the web (we recommend Douglas Baldwin's book and web page or the grand cookbook, Modernist Cuisine—see the bibliography). In the illustrative table we have connected the various heating techniques to the various forms of heat and given each its own icon.

Important heat transfer mechanisms during cooking

Type of heat example	Types of heat transfer			
	Conduction	Convection	Radiation	Condensation
Pan frying (in skillet)				
Oven baking	 Contact with baking tray			
Boiling (water) or deep frying (oil)				
Steaming				
Barbecuing	 Contact with grate			
Microwaving				
Cooking pit (Chapter 12)				 Water from the food evaporates and condenses back on the food

^ The table shows the different mechanisms for transfer of thermal energy, be it frying, deep frying/boiling, baking, grilling, microwaving, or using a cooking pit.

10

Cold Skillet—Juicy Fish

Tatu the chef came back from his summer holiday with great news: now he was positive that fish should be placed in a cold, and not a pre-heated, skillet. "I have tested it throughout the summer. When the fish is heated along with the skillet it turns out juicier and more succulent compared with if it is put in a preheated skillet." His observation was thus quite the opposite as compared to the instruction generally given for frying fish. And what a perfect starting point for an experiment!

For the next food workshop, Tatu bought a large salmon fillet and put it into a 10% brine solution half an hour before the experiment was to start. Together with the participants we cut out two thicker (neck) and two thinner (tail) fillets. The pieces were weighed, and their thicknesses were measured in order to ensure the pieces were pair-wise as identical as possible. A kitchen thermometer was inserted in the middle of each of the thick fillets. Both thermometers had been calibrated at room temperature (20°C) and in boiling water (100°C) to ensure that any measured differences were not due to the equipment. In addition, we had a nice laser thermometer to monitor the surface temperature of the skillet.

The stage was set for frying. A cold skillet with one thick and one thin fillet was placed on a preheated hotplate. After 120 seconds the temperature reached 160°C, whereby the two other fillets were placed in the now hot skillet. All four fillets were fried until the temperature inside the respective thicker piece reached 45°C, after which that pair was transferred to a separate plate, still monitoring the temperature for a few minutes. While Tatu was master of frying, two of the participants were in control of the measurements. Every 30th second, two participants monitored the temperatures in the thick fillets while two others documented the skillet surface temperature. The fifth participant planned how the measurements could best be visualized in a spreadsheet, and the rest watched eagerly. Not the most efficient use of manpower but fun it was. The temperature curves of the two versions of fish are shown in the diagram.

The pink line in the diagram shows the temperature curve of the "hot skillet" version and the blue line shows the temperature curve of the fish placed in the cold skillet. It is worth noting that the temperature continues to increase for a short while even after the pieces were removed from the skillet. We were not surprised that the curves ended up to be different, but they were opposite from what we had anticipated. To begin with, the temperature increase was highest in the fish placed in cold skillet, while evening out toward the end, and the two fillets ended up close to the same final temperature at around the same time. Weird and unexpected!

As usual in our workshops, we had to turn to discussions and review the literature to try to understand the results. We came up with three possible explanations of these counterintuitive graphs, presented below in the order of decreasing discomfort and fear:

1. Looking at the graphs, the skeptic would say that our rather relaxed experimenting group mixed the samples. Actually, we did not! We were under the close scrutiny of many critical eyes, and in this group much smaller mistakes than this are usually noticed.
2. Perhaps our thermometers were not working properly? Well, we did have quite ordinary kitchen thermometers, and the calibration procedure would not exactly stand up to scientific standards. It may also be that we did not succeed in placing the two thermometers exactly into the center of the fish samples. Definitely a weakness that asks for repeating the experiment. We are humble enough to accept this criticism as we did not bring another fillet for a second experiment. So, please go ahead and repeat our experiment!

3. Reading current literature of the field, however, indicates that there might be another explanation, and that the counterintuitive result might yet represent the actual state of affairs. In the great encyclopedias of modern cooking, *Modernist Cuisine*, we found a claim that placing meat onto a hot skillet might lead to formation of an insulating layer. The assumption for our subsequent reasoning would be that the same should apply to fish.

Fish frying experiment

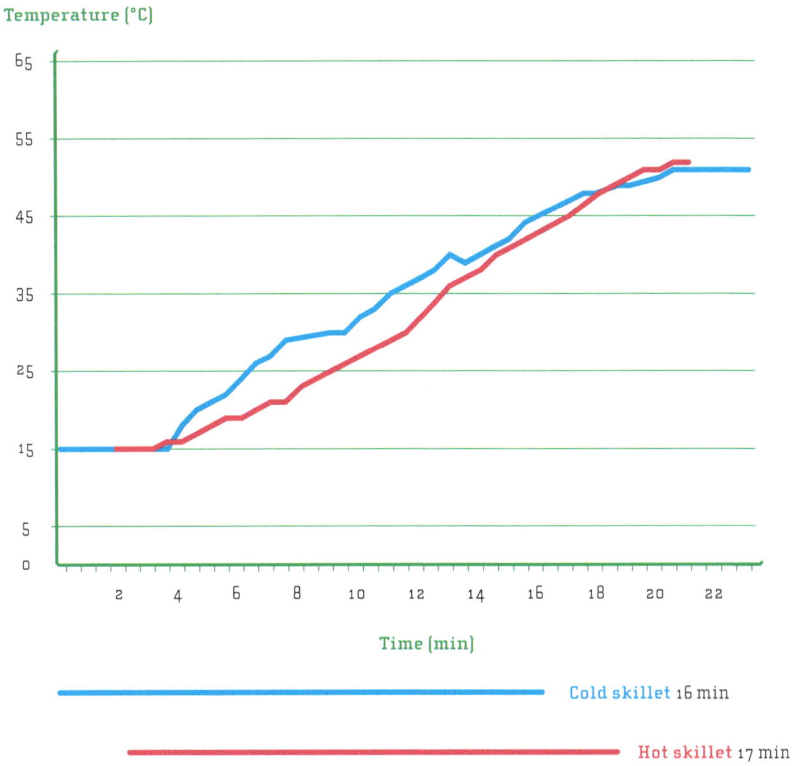

^ The diagram shows the development of internal temperatures during frying. The fish placed in hot skillet actually took longer to reach the desired internal temperature (pink graph).

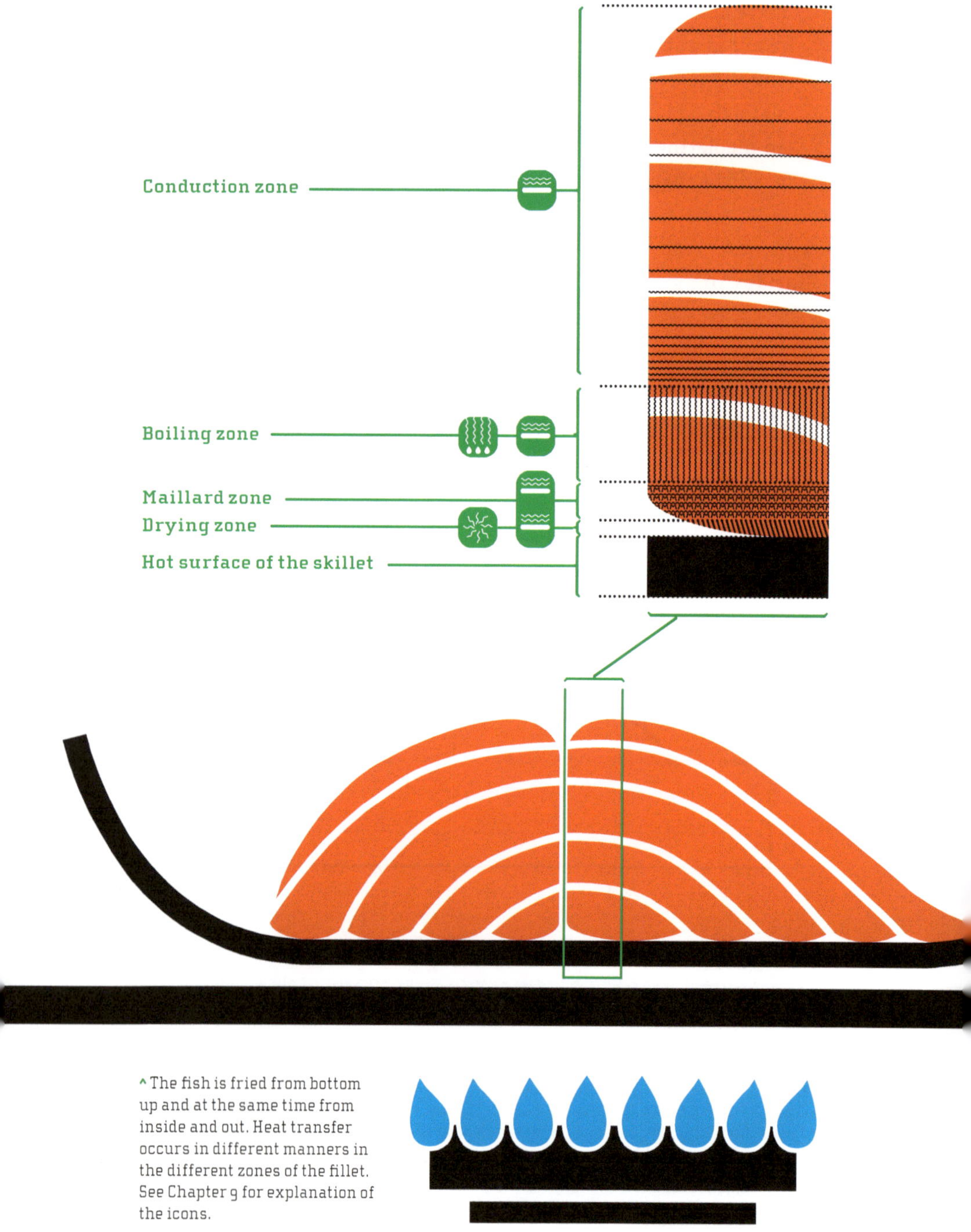

^ The fish is fried from bottom up and at the same time from inside and out. Heat transfer occurs in different manners in the different zones of the fillet. See Chapter 9 for explanation of the icons.

The main author of *Modernist Cuisine*, Nathan Myhrvold, writes that a steak on the skillet, to a certain extent, is cooked from the inside out. When a steak is placed on a very hot surface, four different horizontal zones are rapidly formed from the surface touching the skillet surface and inwards: the zone closest to the skillet, the "drying zone," is a very thin layer between the hot surface (approximately 200°C) and the food. Here, the temperature increases rapidly to pass the boiling point of water, resulting in the water evaporating quickly to give a dry surface or crust. Above (just inside) the drying zone the temperature increases to at least 140°C, enabling the Maillard browning reactions to occur (see info box). This "Maillard zone" receives heat by conduction from the drying zone. Since the temperature in this zone can reach up to ca. 150°C, water boils, rendering this zone dry as well. The next zone going inwards in the meat is a "boiling zone," where hot water continuously evaporates and condenses, maintaining a fairly constant temperature around 100°C, and pushing steam both inwards and outwards. The steam moves mostly from below and upwards through existing and newly formed channels inside the food. Fats melt, connective tissues consisting of the protein collagen are transformed to gelatin, which can dissolve in the water. Meat proteins

denature and coagulate. The meat is at the same time fried from the outside and boiled and steamed from the inside. As the steam forms new channels within the meat, the muscle fibers detach from each other breaking down the chewy structures.

Boiling of water takes a lot of energy and, therefore, the boiling zone is not very thick. Further in is the "conduction zone," where heat principally moves via conduction as well as convection through channels in the meat. In the illustration we have envisaged that the same happens in a fish fillet as in a piece of meat.

The heat conductivity of meat and fish is rather low (see the previous chapter), and thus heat transfer by conduction occurs relatively slowly. In our study, it took 16 and 17 minutes, respectively, for the internal temperature of the fish to reach 45°C. It took a shorter time for the fish placed in the cold skillet to cook compared to the hot pan technique, but weighing revealed that both lost approximately the same amount of liquid: 11% for the cold skillet fillet, 12% for the one placed in hot skillet. The appearance of both versions was golden brown with nice crisp crust. Although the measured difference and external appearance was too marginal to indicate a difference between the two methods, again we realized that the proof of the pudding indeed lies in the tasting. All nine participants reported that

The Maillard reaction contributes to develop both color and flavor and is among the most important reactions in cooking. The reaction, which is in fact several parallel chemical reactions, has been named after the biochemist Louis Camille Maillard. In 1913, he published his findings on color and aroma changes when a mixture of amino acids and carbohydrates are heated in a test tube. The details of these complex reactions were not revealed until 1953, when the chemist John E. Hodge described the basic steps of the reaction. According to his descriptions, a complex multistep chain of reactions take place to produce a large number of color and aroma compounds, thereby creating the distinctive color and aroma in many of our heat-treated foods such as fried and roasted meat and fish, bread, roasted coffee, and malt for brewing and chocolate. ×

they would prefer the fish placed in the cold skillet to that placed in the preheated skillet.

Indeed, the hot skillet version seemed to have several indications of overcooking: more white coagulant, thicker and darker crust. The thicker gray, "overcooked" layer and higher amount of coagulated protein between the muscle filaments indicates that this fish has reached a higher maximum temperature than the cold skillet version. This had happened despite the fact that the core temperature initially increased faster in the cold skillet fillet.

Our kitchen measurements did not meet the accuracy of laboratory measurements, and, thus, we only can guess what was really going on inside the fish during frying. The shorter cooking time for the fish placed in the cold skillet seems counterintuitive. But taken that the result is correct, we have a possible explanation. In a warm skillet, the surface of the fish will dry out quickly to give a dry, insulating layer. This drying zone retards heat transfer into the fillet and can increase the cooking time and give a thicker overcooked outer layer. Secondly, the boiling zone might play a role as longer cooking times should result in the boiling zone, with its high heat intensity, to be thicker (see illustration). Since we aim at an internal temperature far below 100°C, a thick layer at this temperature is not what we seek, as it would overcook the internal parts as well.

In the subsequent discussion, seven out of the nine participants said that they would hereafter place the fish in a cold skillet. The remaining two panelists stating that they would stay with the classic hot pan technique were not unmoved, though, and said that they would go for lower starting temperatures in the future.

Is it time to forget old advice on how to know when the skillet is hot enough to put the fish in and start frying? Such difficult-to-remember advice can be:

- Adjust the stove to medium heat and allow the skillet to heat up. To check the temperature, add a little water. If the water drops bounce and evaporate quickly the pan is hot enough.
- Make sure the skillet is heated to the right temperature. Too hot a skillet burns the fish but, too cold a skillet and the fish will not get the right color and crust.

- Heat butter or margarine in a sufficiently large skillet. When the fat has stopped foaming, add the fish fillets and reduce the heat slightly.
- Add a splash of oil in a dry skillet and heat it until very hot. Place two of the pieces of fish in the skillet with the skin side down first, add the other pieces after few moments. This way you avoid the skillet cooling down too much.

If our experiment is indeed correct and reproducible, you can now forget all these and other similar rules and replace them with the very simple:

- Put the fish fillets in a cold skillet and turn the heat on.

Cold-skillet Salmon á la Tatu Lehtovaara

Ingredients

150 g quality salmon fillet
1.5 g salt (ca. 1% of the weight of fish)
5 g cold butter
Black pepper

Procedure

Sprinkle the salt evenly onto the fish fillet and let it rest in the refrigerator for 30 minutes. Tap dry the fish carefully, rub the cold butter onto the skillet and place the salmon fillet in it meat down. Move the skillet to the stovetop and turn the heat on a little over half power. Fry the fish until the internal temperature is ca. 35°C, turn the fillet and continue frying until the internal temperature is ca. 45°C. Season with pepper and remove from skillet.

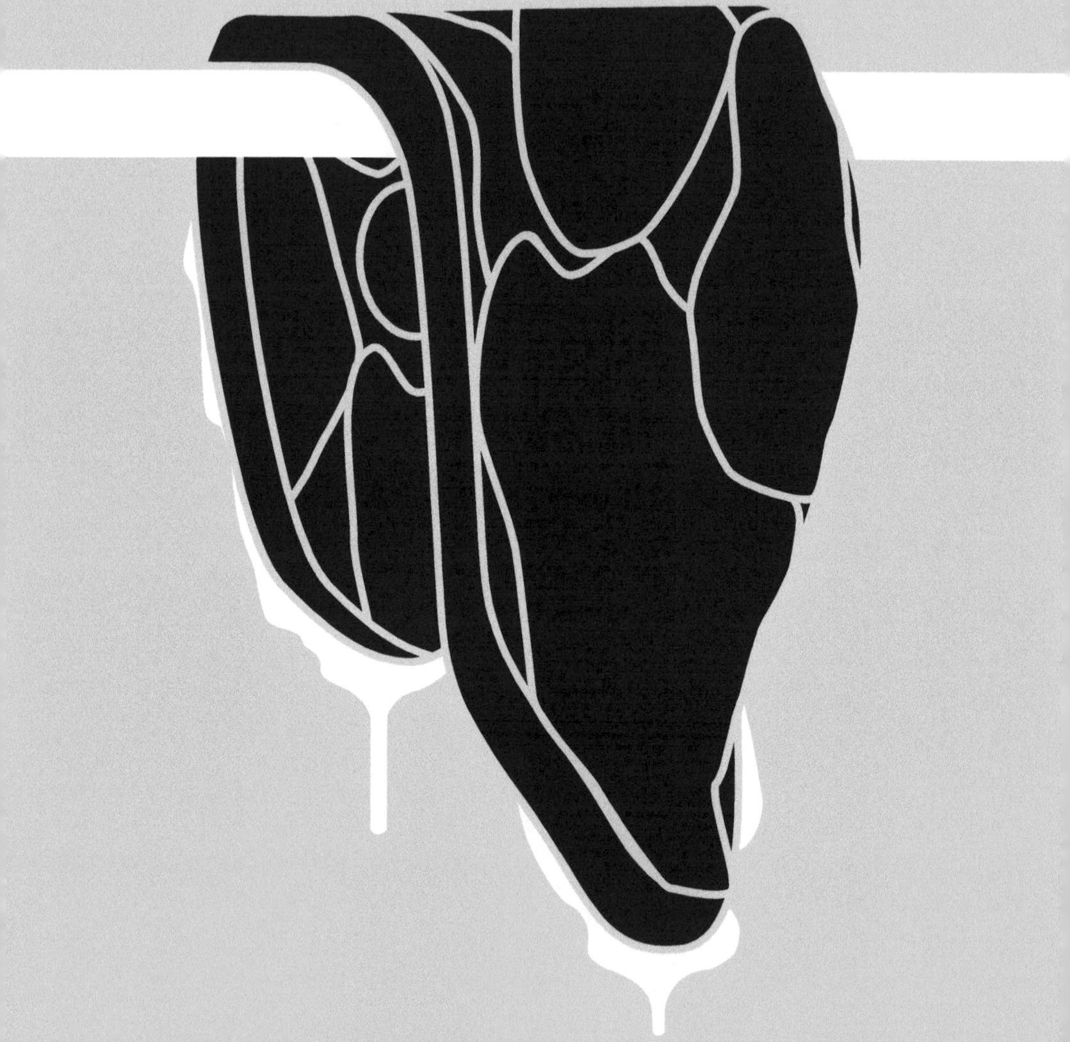

11

Tempering the Meat Before Cooking, ...Or Maybe Not

One day chef Tatu presented us with a rather surprising question: "Is it really necessary to temper meat before cooking?" A surprising question coming from a chef because the answer is so obvious: yes, of course! Everyone surely knows that we should take the meat out of the fridge well in time before cooking, be it a roast or a steak. Indeed, this advice is written into every chef's DNA, and it is simple common sense that meat will be more evenly cooked if the meat is not cold inside when placed in the oven, skillet, or pan. Cooking a piece of meat straight from the fridge should result in a thick outer layer of overcooked and drier meat before the center reaches the required temperature of, say, 55°C. Cold meat loses more juice during cooking than a properly tempered piece of meat. Or will it...? If this is not true after all, it could indeed be somewhat of a culinary bombshell.

But we have been surprised before, such as when Tatu asked if we really had to preheat the pan before frying fish (in the previous chapter). These types of statements are the most interesting ones, as we really need to challenge our own biases and preconceptions. As we have often experienced, there are factors we hadn't thought of coming into play, showing us that the intuitive and linear way of thinking may often not be sufficient. The question called for a carefully designed experiment.

The cold meat experiment. We set up an experiment with two parallels to test if we could taste, or measure, any difference in pre-tempered meat and that taken straight from the fridge. We bought two pieces of beef tenderloin and cut both into two identical samples for each parallel. Duplicate samples were used to compensate for any differences between the pieces, one thicker (6 cm) and one thinner (5 cm) in each parallel. This should give a certain reproducibility; although, the experiment was not as rigorous as a proper scientific experiment.

All fillets were wrapped, and while two were put in the fridge, the other two were left in room temperature. A couple of hours later, we inserted cooking thermometers into the fillets and placed them together in the oven at 100°C. The tempered samples started at 17°C whereas the fridge cold ones started at 6°C. Each fillet was taken out of the oven when it had reached an internal temperature of 55°C. Thus, the cooking times were different, but the degree of cooking was the same for all fillets. After cooling, the fillets were weighed, carved, and blind tasted by 15 food workshop participants.

The participants were asked to select which of the four samples was the juiciest, the driest, the most red, and the most grey. And finally, they were asked to select which was the most and least preferred.

Surprising results ahead! In the blind tasting, the fridge cold fillets were evaluated as both most juicy and most red, and the room tempered ones were experienced as the most grey and dry (see the diagram). The more subjective votes for liking also went in favor of the fridge cold ones. Had we really falsified the claim that meat should be tempered before cooking? Such a controversial result needed more backing, and so it got from the more objective measures of cooking times and weight loss, or, in food terminology, cooking loss. The cooking times were 29 and 40 minutes for the room tempered ones, and 35 and 43 minutes for the fridge cold ones. The thin ones, as expected, were those requiring the shortest cooking time within each pair. The

Blind tasting of tenderloin fillets

^ The diagram shows the results from blind tasting. The panel evaluated juiciness, color, and personal preferences. The results were pretty surprising!

weighing indicated that the cooking loss was highest in the two tempered meat samples. They lost 14.1% (thinner) and 18.0% (thicker) of their weight during cooking whereas the cold meats lost 14.1% and 13.6%, respectively. The thick and cold piece of meat thus lost less of its weight during cooking. Allow us to repeat: the thickest of the fridge cold fillets had the lowest cooking loss of them all! To our knowledge, the direct opposite of what common kitchen wisdom would tell us. Of course, for proper research, we would need to repeat these experiments several times, calculate average values and do the statistics. It is worth reminding ourselves that the tasting result is partially a result according to today's preferences, and preferences may change over time. If we had carried out the blind tasting 50 or 100 years ago, when many of our cooking rules were established, the votes for liking and thus cooking ideals might have been different.

Standard cooking rules, expert advice and common sense would tell us that the thickest and coldest fillet would be the one that lost the most water. Both because it cooked for a longer time and because the outer layers should expectedly become overheated for longer before the center reached the final temperature. After the first shock of being wrong we noticed that being wrong is a great way to learn more, keeping us on our toes at all times. Should we consider changing the instructions? Maybe not quite yet, but our results did indicate that this was a topic for further research. Our experimental setup was rather simple, as we used only one type of meat and one way of cooking. However, the unexpected results that we got both from objective weighing measurements and subjective sensory evaluation forced us to go into the literature. Interestingly, we didn't find research on this.

So, we sat down to think over how heat spreads in meat. Unfortunately, we did not have fancy thermometers that could measure temperatures in different parts of the meat. However, we did have access to theoretical models on heat transfer in food during cooking such as the "Cook my meat calculator" developed by MIT students during a scientific cooking class in 2013. According to this model (see figure), the outer layers of tempered meat should heat up more rapidly than in the fridge cold meat. Throughout the whole cooking process, the room tempered meat would, according to the mathematical model, be warmer than the fridge cold meat, the latter catching up in the end. In the end, when the internal temperature had reached the targeted 55°C, the temperature distribution in both fillets were identical. So, they eventually ended up at the same final temperature, but the route to this temperature would not be the same. They would take somewhat different time, with a theoretically different "heating history" and the outer layers of the tempered meat would experience higher temperatures for a longer period of time compared to fridge cold meat.

Taking it one, or two, steps further. This should be good news. If the well-known cooking rule of tempering the meat is indeed incorrect, there would be a threefold benefit. You save time, the meat may come out juicier and you avoid leaving your meat at temperatures where microorganisms in the surface might multiply.

We didn't test this for larger cuts, such as a whole leg of lamb or a whole ham, but the results brings to mind the traditional Finnish dish "frozen roast." This is an exceptionally tender and juicy roast made by placing a frozen roast, preferably game such as moose or reindeer, in a 90–100°C oven and cook it overnight. The meat first thaws, and

Heating of fridge cold and tempered meat at 100°C

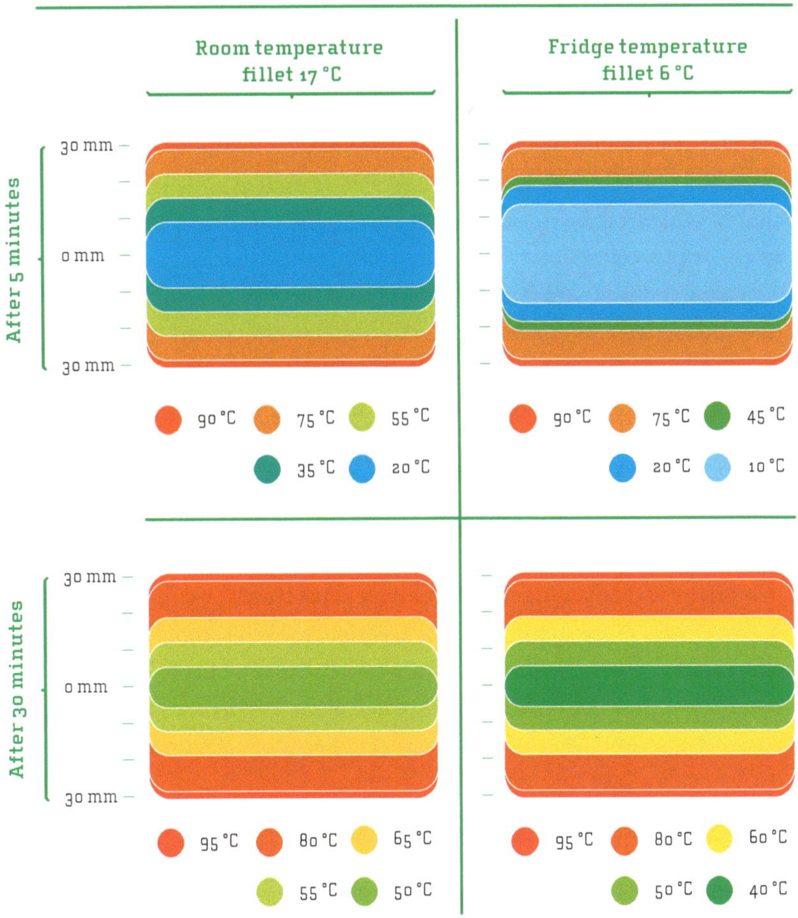

^ Theoretically calculated temperature distribution in fridge cold and tempered meat during cooking at 100°C. During heating to internal temperature 55°C (not shown), the fillets end up with a nearly identical temperature distribution. Illustration based on calculations using the "Cook my meat calculator" developed by students at MIT in Boston.

then is cooked slowly at low temperature. In the morning the roast is taken from the oven and immersed for several hours into a slightly salty stock made beforehand by simmering salt, spices and herbs in water. The meat absorbs the salty, aromatic water to give a tender and juicy piece of meat full of flavor. Apparently, cooking cold meat is not a new invention—we have just forgotten how good it can be.

However, our experiment did not tell us anything about how the meat reacts if we want to fry it in a skillet like a steak. Our method of cooking the meat wrapped in plastic at 100°C is enough to make it reach the desired internal temperature, but not enough to let the Maillard reactions give a brown and tasty crust. In a later workshop we tested this and observed that meat placed in the skillet directly from the fridge develops a crust and color that is at least as nice as that of tempered meat. Actually, it was slightly darker and more even in the fridge cold fillet.

Apparently, the cold meat did not cool the skillet as often claimed (we made sure not to overload the skillet with meat). Maybe this reflects a general thermodynamic fact: the energy required to increase the temperature in water—the main component of meat—the 15–20°C differentiating tempered and fridge cold meat is negligible compared to the energy required to evaporate the already hot water in the meat, thus reaching temperatures allowing the Maillard browning reactions to occur. The more even color of the crust in the meat fried directly from the fridge may arise from the cold meat being more rigid than the tempered, allowing for fewer and smaller pockets of steam between the skillet surface and the meat. Indeed, chefs have tried frying steaks straight from the freezer reporting good results. Some of these are described on the Internet, among others in a video by America's Test Kitchen.

We gladly agree that our tests are incomplete and require further experimenting before we would entirely dismiss such a long-lasting cooking rule as the one about tempering meat. So, we encourage everyone—amateurs, food professionals, and academics—to seize the opportunity and start experimenting with a systematic and open mind!

12

Cooking Pits—Prehistoric Cooking Meets Science

Those were the days—a historical perspective. Let's wind the clock back from our meat and fish cooking experiments to the period around 3000 years before our food workshops. We are reaching back to the age that, in the Nordic countries, is called the Bronze Age, around 1750–500 BC. The inhabitants in this region had long since become more stationary, living at more permanent sites rather than moving around to follow the food as they did during the Stone Age (it is worth noting that historic timelines vary between regions of the world. The now Nordic countries were inhabited fairly late since animals and humans followed along as the ice withdrew northwards, and the last ice age ebbed out roughly around 13,000–10,000 BC). In any case, these are prehistoric times meaning that we don't have written material to guide us. Only ancient remains of people's lives are available for archeologists to interpret by piecing them together as logical jigsaw puzzles to construct a picture of how our ancestors lived their lives and cooked their food. One such puzzle is the uncovering of shallow holes in the ground containing remains of coal, burnt and cracked rocks, and microscopic amounts of animal fat. Surely there must have been cooking going on in these? These holes, commonly called cooking pits or earth ovens, can be dated using carbon dating, which has revealed that they were in use in the Nordic countries already in the Paleolithic Era, but reaching their peak during the Bronze Age, and staying in use until late in the Iron Age. Such pits have been found basically all over the world and are still in use some areas with their respective characteristic patterns in terms of the depth, size, and composition of the organic material depending on what was cooked, and the local practices employed. Indeed, cooking pits are still used today, but here we will turn to the prehistoric ones on the Nordic countries.

In Europe, they seem to be common from Central Europe all the way up to Lapland (the region called Sápmi, stretching across northern Norway, Sweden, Finland, and into Russia). In the 1960–70s, archeologists started using motorized machinery for mechanical topsoil stripping. This way, patterns across larger areas of land could be mapped out. Rather than the odd single pit from the Paleolithicum, pits were shown to appear in fields close to each other, from 20 up to so many as 500. Such collections of pits are found close to areas where people had lived after establishing themselves at more permanent locations. By coordinating this information with other findings in the same area, archeologists get a picture of how the pits had been used and even make hypotheses of the original purpose of the pits. Were they used for daily cooking or for special occasions only? Was it food that was cooked, or were they graves where dead people had been burnt? Perhaps here had originally been a small tent on top of the pit that had long since decayed? In that case, they may have served as saunas, as drying or smoking houses for meat or fish, for preparing hides, or as drying houses for grains.

Whenever an archeologist makes a finding, s/he will have to interpret the data just as when we do our tasting experiments. Almost always there will be more than one possible explanation to explain the origins of the finding. The combination of rocks, coal, and fat indicate quite clearly that they were used for animal biological material. Secondly, they haven't found human bones in the pits, which of course is no direct evidence that humans were not burnt there because our ancestors could have removed the bones after the burning ritual. So, the interpretation of the pits as some sort of "boneless tombs" is not a

Historical timeline in the Nordic countries

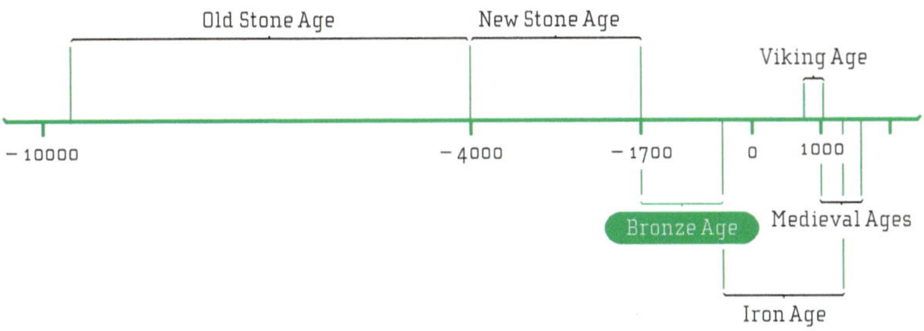

^ **The age of the cooking pits:** Most of the cooking pit fields found in the Nordic countries are dated to the Bronze Age. The timelines in different parts of the world vary depending on the cultural and technological development of the region.

hypothesis that can ever be falsified, and thus the hypothesis is considered by archeologists to be somewhat speculative. After all, as the science philosopher Karl Popper wrote (in our words): every suggestion for an explanation should be testable, and if there is no way to test if your hypothesis is false, then your hypothesis has a serious problem. So, we cannot rule out that it actually happened, but there is no way to test if that interpretation is wrong either. We can choose to leave this as a possible use for cooking pits, but we will never be able to confirm or falsify it. Frustrating, but this is what researchers must learn to live with from time to time. Were the pits used for "dry cooking," or were they lined and filled with water to be boiled by hot rocks? Plunging hot rocks into water results in different cracking patterns than burning them dry. This is a fortunate fact because the cracking pattern of the rocks found in the pits can then be used as evidence to show that water was not used in these pits. So, wouldn't the finding of dry-cracked rocks also serve to disprove the sauna hypothesis, as it also disproves wet-cooking? As we envisage possible uses for these structures in the ground, new hypotheses appear that must be evaluated, disproved or left as possible explanations. But let us, for now, with our eyes open for other possible explanations, choose to see these pits as actual cooking pits used for cooking food for some occasion. After all, we won't be criticized too heavily by archeologists for this choice as this seems to be the prevailing hypothesis among many of them.

The next question is: did the prehistoric Norsemen and -women use these subterranean ovens for everyday cooking for the whole family or village, or was this reserved for special occasions only? When cooking pits went out of use during the late Iron Age, was this because we invented new materials and new technologies or was it for some other reason? Archeologists putting their interpretive minds to the task, piecing all the various bits of evidence together, tell us that cooking in a pit was probably not an everyday activity; the village did not come together to cook a whole venison every other day. Excavations reveal that the pits were not reused many times, and the number of pits in an area compared with the other remains of human activity tells us that the number of pits is far too low to be a regular method for feeding the people living there. Even if the people had meat or fish only once a week. There are reasons to believe that cooking pits were used for religious purposes such as offerings in pre-Christian religious practices: *blót*, in old Norse language. There are, for example, remains of horse meat found in and around cooking pits near large houses, which were centers of power and homes of important chieftains. The horse had a strong symbolic meaning. Finding horse meat in a cooking pit might be related to offerings of symbolic character to demonstrate power, hierarchy, and wealth, but also a wish for good fortune in battle and agriculture. So, apparently, there are other possible explanations for why people used cooking pits than plain nutrition in a low-tech society compared with ours.

In recent years, pit cooking has regained some popularity as both a spectacular, fascinating but also effective way to cook food for everything between five and 150 people. Indeed, it is not much more labor to cook for 50 than for five. Those of us taking the effort to dig a hole in the ground, collect rocks, kindle a fire, and wait for the hours it takes are surely seeking an experience out of the ordinary. The uniquely complex smell of cooked meat, aromatic herbs, and boiled soil reaching you the moment you lift the turf to uncover the pit is an experience that may leave its olfactory imprint on you for years. Cooking in pits are also, today, inevitably something for the special occasions rather than an everyday activity.

So, is the increasing popularity of cooking in pits also of a more symbolic and experience-oriented nature, some sort of mild "food cult?" After all, those of us who try out pit cooking might seek some extraordinary experience or a unique flavor. Or do we do it to distinguish ourselves from the everyday cooking or ready-made food? One thing is certain: the cooking pit experience is a unique and recommendable one, regardless of our motivation to engage in it. Indeed, the Norwegian archeologist Lars Erik Narmo has written that such "dry-cooked" meat from a pit turns out very juicy. Upon such a statement, we, as food-oriented chemists, feel that there are some gaps to be filled: why does the food turn out juicy? Which kinds of cooking processes take place in a cooking pit? What is, if any, the difference between pit cooking and more modern methods for cooking? Do we really need to dig a hole in the ground to make this delicious food, or can we accomplish the same without engaging in a prehistoric role-play (unless we find this meaningful or fun in itself)?

Making a cooking pit. The Northern European Bronze Age pit chef would most likely look for a flat space with fairly dry soil, not sand, and good turf that could be cut in whole pieces. S/he would then

Procedure for preparation of food in a cooking pit

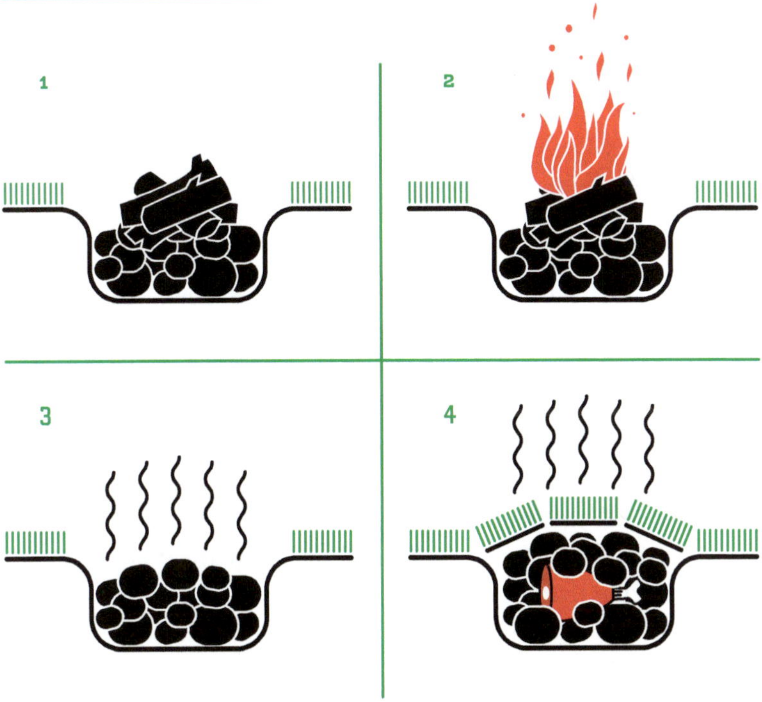

dig a round, oval, or quadrangle hole of 20–40 cm depth and from 1–3 m diameter in width, being careful that the turf was left in whole pieces and put aside. While lighting a fire in the pit, the party would collect round rocks the size of a hand, or somewhat larger, and place them on top of the burning fire. While the fire was burning, the meat would be wrapped in some material, perhaps a hide, animal stomach, birch bark, large leaves, or clay. What they used for wrapping is difficult to trace because most of these materials have long since decayed leaving no trace for the archeologists (we have not found literature stating the finding of clay with traces of food in relation to cooking pits). As soon as the fire had burnt down, the food was placed on top of the very hot rocks and coals, and the turf rolled back to cover it all. You now have a pit containing a layer of hot rocks, food on top and then turf as lid, a literal earth oven. No live fire, only hot rocks to heat the food. One skill of the expert prehistoric chef would perhaps be to know exactly how much rock was needed to heat a certain amount of a certain type of meat to perfection. Surely, if you were cooking a salmon, it would require shorter time and less rock than a whole leg of a reindeer.

It is not difficult to make your own cooking pit as long as you have the available land area, are allowed by the authorities to light a fire and have access to rocks. Archeologist Lars Erik Narmo writes that the method described above works fine, and there is no need to put rocks on top of the food. This is very similar to what in Finland and Sweden is called "robber's roast." Maybe the Norwegians were law-abiding people in old times, because we haven't found any similar name for this dish in Norwegian. The food is packed in aluminum foil, wet newspapers and chicken wire. After the fire has burnt down, the coals are removed, the food placed on top of the stones, coals shoveled back on top of the food and finally covered by turf or sand. In our own cooking pit practice, which we have done with our students in Norway every year for more than a decade, we dig the hole deeper, use more rocks compared with what described in the literature, and we place the hot rocks all around the food to ensure even heating from all sides. Does this make a difference compared with the prehistoric way? We really don't know, although Narmo, without giving further details, writes that "experiments show that it is not necessary to place rocks on top of the meat." If you want to have rocks both below and on top of the food, your need to extract some of the hot rocks from the fire, place the food on the bottom rocks and put the smaller rocks back on top. This requires thick, heat-resistant gloves or some tool to handle the hot rocks. It is not easy to estimate how much rocks you need to collect, but our years of experience in pit cooking lambs legs have taught us that you probably need to collect about double the amount than you initially would think is enough. When cooking whole fish such as salmon, much less rock is needed as cooking time is shorter and the required temperature is much lower. Usually we would cook both a leg of lamb and one or two whole salmons, and some vegetables. Can you manage with one pit or do you need two? As we have the manpower, we usually dig one for the lamb and one for the fish and start the fish later. The question then remains where to place the vegetables, which will always result in a compromise in terms of cooking time. Surely there is always place for further experimenting.

Procedures on the web indicate different cooking times, but when we cook lamb leg, it usually requires 2–2.5 hours, with 3–3.5 hours to give the best result. When using longer cooking times, the bone can simply be pulled out to leave the now very tender meat in one piece. Increasing the cooking time will not necessarily result in a very dry meat as when using an electric oven because the pit will gradually become cooler the longer you wait. When you are finished eating, make sure to put the rocks back where you found them and fill the pit with the soil and turf so that there are as few traces as possible from your activity. Archeologists recommend that you drop a small coin into the pit before filling it back up to make sure that later excavations don't confuse your pit with a prehistoric one, spending time and resources on carbon dating material from a family party from the 2000s.

Pit cooking—gradually turning down the heat. Cooking in a pit is quite different from what most of us do at home or in restaurants in more than one sense. Obviously, we don't go digging holes in the ground every time we want to cook a roast, leg, or fish. But pit cooking is different from most cooking in the respect that when the food is placed in the cooking vessel or oven, the heat source has already been turned off. Ordinary oven cooking is usually conducted at a constant oven temperature, cooking in water or gravy/sauce is done at constant 100°C or lower and frying an egg in a skillet is done while the hotplate constantly feeds energy to the skillet. In the cooking pit, however, the fire has gone out before the food is added. While the firewood is burning, some of the thermal energy released in the process is transferred as heat to the rocks, which in turn become hot. The rocks have a certain *heat capacity*, ability to store thermal energy, and when the pit is closed we have a starting situation for the cooking process where the cold uncooked food is surrounded by hot rocks. Hot and cold objects in vicinity or contact with each other will engage in energy exchange. The hot becomes gradually colder by giving off excess energy, the cold surroundings, which include the food, become hotter by receiving this energy as various forms of heat. Through this exchange of energy, if we wait long enough, we ultimately end up with an equilibrium where everything has the same temperature. The ability to store thermal energy is a property of the material that can be measured, and for a specific material this is called the material's *specific heat capacity*: the amount of energy it takes to heat 1 kg of the material one degree Celsius (or Kelvin). Materials with high heat capacity take more energy to heat and can therefore store more energy.

When we pile food and hot rocks together in a cooking pit and cover it with the turf, we ensure that more of the thermal energy in the rocks is transferred to the food rather than to the surrounding air. In a cooking pit, all four kinds of heat are in action:

- conduction through direct contact between rocks and food.
- convection of hot air and steam inside the pit and inside the packages of food.
- radiation from the hot rocks reaching nearby food.
- condensation of water vapor, both on the surface of the food packets (humidity from the soil) and inside the packets (water from the food).

Effectively, pit cooking combines almost all forms of heat treatment: boiling, steaming, barbecuing, oven roasting, and skillet frying in one single cooking method.

As shown in the diagram, water has a very high heat capacity; large amounts of energy are needed to heat it up. Since most food contains a large proportion of water, this means that we need relatively large amounts of rock compared to food. This is, however, compensated by the fact that the rocks are very hot to begin with. Heat

Approximate specific heat capacity for various materials

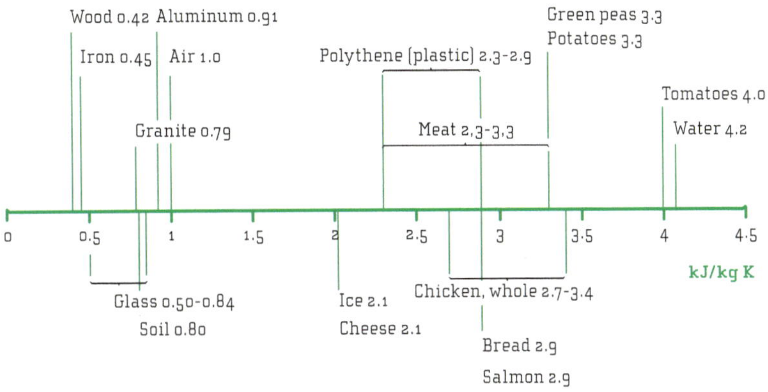

^ Foods have varying proportions of water, which contribute considerably to the heat capacity. The unit kJ/kg K is kilojoules per kilogram and Kelvin. That is, the energy it takes to heat one kilogram of the material one degree Celsius (or one Kelvin).

transfer is, after all, a product of heat capacity, amount of material and temperature difference. Additionally, we must take into account that much thermal energy is lost to the surroundings, that the food has to maintain a certain temperature for a certain time to be cooked and so forth. So, as result of both the energy loss to the surroundings and the difference in heat capacity between rocks and food, it is a good idea to be generous when collecting rocks for a cooking pit.

Tender meat from the pit. When the archeologist states that food that has been "dry cooked" comes out very juicy, he particularly refers to a statement from a huntsman pit cooking a leg of moose. Having very large legs and very lean meat, moose legs would easily result in a dry chewing experience. Cooking a whole leg of moose might indeed be the ultimate test to show if pit cooking can compete with other methods. Two aspects are often emphasized for good taste and tender meat:

1. High starting temperature. The surface of the meat is seared at a high temperature, which promotes Maillard browning reactions that produce tasteful compounds and color. This occurs in the early phase of cooking in a pit when the rocks are very hot (the rocks may be as hot as 500°C or higher).

2. Moderate or low temperature over a longer period of time. For the meat to stay juicy it should be given sufficient time at moderate temperature so that tough connective tissue (collagen) is transformed to soluble and tender gelatin. Also, other chemical processes occur to produce flavor compounds. Although rather poorly controlled, the temperature in a cooking pit is constantly decreasing, resulting in an increasingly milder cooking.

During cooking, the meat has been tightly wrapped only lying in its own juices. This should prevent severe overheating as well as stop moisture from escaping. If the guests should be somewhat delayed, we can be less worried about the food staying too long since the temperature in the pit is constantly decreasing. So, the cooking pit method has at least two safety mechanisms against severe overcooking built into it.

Still, although not intentionally, we have been able to overcook food in a pit. In that case, we were a bit too industrious and used more rock than necessary. Cooking delicate products such as fish has an inherent greater risk for overcooking, and potatoes may occasionally turn out burnt and dry. Opening the pit to check on the food is not a strategy we would recommend because it is quite a lot of work, and when it has been opened, it is not easy to put the lid on, and much heat will have escaped. So, if time allows, we usually cook the food a bit too long rather than too short. Occasionally we use a digital cooking thermometer just to have some clue about how things are going (be prepared that the thermometer probe might be broken in the process). Cooking in a pit is not an accurate science, it is rather an experience-based practice that usually turns out very well. After all, the Bronze Age shaman didn't have a digital thermometer lying around. Today, however, we can do such experiments to test whether our theoretical assumptions can be confirmed by measurement.

What goes on underground? When cooking with students, we usually monitor the temperature inside the food. But we also track the temperature in various places in the pit, including in the ground just outside the pit. This way we can get some hints to what goes on, how a cooking pit works and if much energy is lost to the surroundings.

Lamb leg in cooking pit

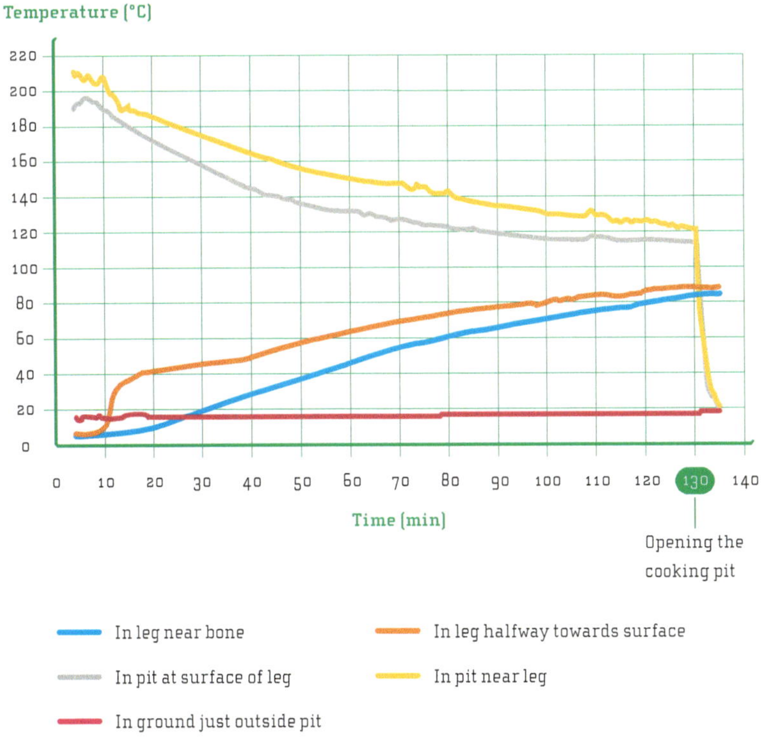

^ **Temperature zones in a cooking pit:** The graphs show temperature development measured during cooking of a leg of lamb. The temperature was measured in the meat, at the surface of the meat, two places inside the pit and in the ground just outside the pit.

Our assumptions about the temperature development in the pit and the food seem to be well supported by our measurements. As seen from the temperature in the ground just a few centimeters outside the pit, it seems well insulated and not much of the heat is lost to the immediate surrounding ground. This cooking experiment lasted just above two hours, which gave a well-done leg of lamb based on the appearance and texture with a brown seared surface. Although not particularly dry, it was not very juicy either. It was less dry, however, than expected from meat cooked to 86–90°C, as the graph shows. Interestingly, the outer and inner parts of the meat are getting closer in temperature as time passed even though the pit was still 30 degrees warmer. A notable result was that the outer part of the meat was rapidly heated in the beginning, but after a while the temperature evened out. Apparently, the meat was able to transfer energy inwards quite efficiently, rather than being much overcooked in the outer parts before being ready along the bone. This is in good agreement with a contemporary experiment described in the chapter about tempering the meat.

It would have been interesting to leave it for another half or full hour, but this time the time constraint of the teaching session did not allow for this. Previous experiments with longer cooking times have shown that the meat indeed does not become drier, but it becomes more tender, falling apart and resulting a pulled-meat structure. Even if the cooking method is sometimes called "dry cooking," the food is so tightly wrapped that a large amount of water from the meat itself cannot escape the food. Dry-cooked meat will end up rather moist. So, our experiences and experiments, although somewhat crude, do indeed support the claims by the archeologist that cooking food in a pit produces a delicious result. However, the question remains: does the food come out better if we spend the better of a day digging, collecting rocks, burning a sack of firewood, and waiting for another few hours? Most likely, it would be possible to achieve a very similar leg of lamb using techniques and tools from your own kitchen. But you would miss out on a great day out and the unique sensory feast that is opening a steaming pit. Perhaps the ultimate distinction from ready-made food produced in a modern factory would be to take on the role of the prehistoric chef, or shaman, from 3,500 years back in time, while at the same time being able to withdraw to the warmth of our living room whenever we want? Isn't that indeed participating in some sort of joyful ritual, going back in time to sit at the table with our ancestors?

›››

Other cooking methods functioning by the same principle as cooking pits:

- Wood-fired stone ovens—in the case where the embers are removed before the food enters the oven.
- Haybox or straw box where you heat your food to the boil and place it in an insulated box to finish the cooking without further heating. An old version of the slow cooker that saved energy/firewood. Modern versions are available as insulated cooking boxes used to cook or complete cooking without fire. This has been known at least since the 19th century, popular during wartime, and was presented at the 1967 World Exhibition in Paris. Under the name "The Self-Acting Norwegian Cooking Apparatus," it even won three gold medals. Today we can still find a recipe for "duvet porridge" on the back of porridge rice boxes in Norway, where a pot of porridge of rice and milk/water is brought to the boil and left in a cold oven or wrapped in either newspapers or a duvet to finish by itself over a period of 3–4 hours.
- Making boiled eggs by boiling a certain amount of water, adding the eggs and taking the pot aside until eggs are ready cooked (see the chapter on cooking eggs). ×

13

White Wine for Cooking Should Always Be Dry, Right?

You are to cook something that requires white wine, for instance a really good risotto, Hollandaise sauce or steamed mussels. But which wine should you choose? "The one that you drink while making it" is often a good, and indeed common, answer. Although enjoyable, the problem with this recommendation is that it doesn't stand up to scientific criteria. The universal rule seems to be that white wine used for cooking should be dry. The classic encyclopedia on food and cooking, Larousse Gastronomique, also supports this:

There are certain general rules on the use of wine in cooking. Whatever its color, the wine must be clean and without harsh, aggressive taste.... White wines used for cooking are usually dry and rather acid...

The 21 kg, five-volume Modernist Cuisine, which otherwise explains the science behind every thinkable and unthinkable aspect of cooking, simply states in its recipes: "White wine (dry)." With no further comment or discussion on the nature of the wine. These are quite authoritative sources on knowledge about food and cooking, and they are usually credible sources when it comes to explaining *why*, rather than just stating *that*. But, in this case, it seems that they just accept a claim about cooking without asking whether it is actually correct or generalizable, and no clear distinctions seem to be made across various dishes. Such claims immediately stir our curiosity. When someone says "do so" or "don't do that," our reflex reaction as curious researchers of kitchen activities is to investigate, and perhaps try the exact opposite just to see what happens. Dry wine for cooking, is that so? In all recipes, always? How come?

Making a quick internet, food book, and recipe search reveals that this claim is widespread; although, all sources aren't that orthodox or strict. The dryness of a white wine is directly related to its level of sweetness, the amount of sugar in the final product. A dry wine is a wine with low degree of sweetness. In this sense, the ultimate dry drink would be plain water because it contains no sugars at all. Dryness in wine is therefore not about its acidity. The amount of residual sugar in the wine is a quantitative measure of how dry or sweet a white wine is; however, wines with the same amount of sugar per liter might appear quite different depending on other characteristics, such as acidity, mouthfeel, oakiness, and so forth.

Although our printed, high-credibility sources do not say anything about the why-issue, the almighty literature has some suggestions. Some would advise against sweet wines because they will add sweetness to the dish; others claim that higher sugar content in the wine might lead to caramelization reactions in the pan. Heating sugar by itself produces caramel-like flavors, while heating sugar in the presence of proteins (from meat, milk, flour etc.) will result in Maillard

Other claims about white wine in cooking:

- The wine should have a relatively high level of acidity.
- The wine should not have an oaked character as it can impart unwanted bitterness.
- The flavor profile of the wine should resemble the flavor profile of the food.
- One should use the same wine in the food as in the glass. ×

reactions to give nutty, roasted, or caramel-like flavors such as those of bread crust, browned meat, fudge, and so forth. Both these reactions also give brown color. These hints are valuable for us in our experiment as we can monitor possible appearance of these characteristics by our sensory evaluation. However, these explanations are not actual explanations but simply claims themselves in need of justification and elaboration. On the other hand, it is quite common in many savory dishes to add some sweetness for the sake of flavor. Many chefs add a pinch of sugar to their risotto, tomato sauce, or stew to balance and fine tune the flavor. Why use a dry wine and then add sugar? Could we simply have uses for a slightly sweeter wine?

A key live source to knowledge on white wines in cooking is Norwegian Guro Helgesdotter Rognså who wrote her doctorate on emulsion sauces, Hollandaise sauce in particular. Hollandaise is a classic French sauce based on butter, egg yolk, and white wine or lemon juice. Working with top chefs during her doctoral work, she collected experience on professionals' choice of white wine for cooking this sauce. She confirms that there is considerable variation and disagreement among chefs on the choice of wine. In her own experiments they tested fairly sweet wines without noticing much of a detectable difference in flavor. This, she says, could be due to the complexity of the dish masking the rather small difference in wine sweetness, an effect called the "mixture suppression effect." Also, the high fat content of the sauce may mask minor flavor differences. The claim that oaked wines would impart bitterness did not come clearly through in her studies, but oaked wines gave a certain spicy note to the sauce.

The stage was set for a comparative investigation in one of our informal food workshops. As usual, to avoid mixing the cards by trying to test several things at a time, we chose to isolate one single claim to investigate: "white wine used in cooking should be dry." We chose a basic risotto as the experimental frame because this can be made as a homogeneous dish without too complex flavor, allowing for ease of comparison in a blind tasting. We picked two wines that were as similar as possible—except for the sugar content, one dry and one semi-dry. To see if it makes a difference whether the dish is cooked with sugar or if a possible difference is solely due to sweetness in taste, we added a third parallel. One risotto made with dry wine, one with semi-dry, and a third where we used the trick of the chefs:

prepare the risotto with dry wine and add a pinch of sugar in the finishing step. The size of this pinch was calculated so that we added sugar equal to the difference between the two wines. Hence, the third parallel should contain the same amount of sugar as the one prepared with the semi-dry wine.

As usual, our instruments of analysis were the senses of our workshop participants applied in a blind tasting. We chose to evaluate the three risottos according to taste, that which is perceived in your mouth (and not nose, such as aroma): saltiness, sweetness, bitterness, acidity, and mouthfeel. Finally, the participants were asked to rate them according to liking/preference. Although sugar/sweetness is the main issue, other taste sensations may also be affected by the level of sweetness, as we described in detail in another chapter. The three quite delicious risottos were ranked from "most" to "least" for the various taste characteristics.

The main experience from the tasting was that it was quite difficult to judge the differences between the three parallels. In more than one characteristic, the differences were so marginal that we weren't certain whether we really could taste the difference or if we had to guess, and we really had to concentrate very hard to notice differences. When differences are small, results become closer to random, or at the worst they might end up skewed when the number of participants is low. Indeed, an important result was that there was not a clear difference in sweetness between the risottos made from the dry and semi-dry wines. And this was in a context where the participants were really concentrated on looking (tasting) for differences, not a dinner or restaurant table with convivial conversation and other distractions to take focus away from small variations in a dish. This corresponds well with the findings by Guro Rognså, who stated that other characteristics of the wine might be just as important as how dry it is.

One result that many in the panel agreed upon was that adding sugar to the dry wine risotto, equal to the difference between the dry and semi-dry wine, did give a risotto that was experienced as somewhat sweeter than the other two. A possible explanation for this is that we used table sugar, sucrose, whereas most of the sugar from the grapes is glucose. Sucrose is comparably sweeter tasting than glucose, so even though the risotto did not have more sugar per gram than the one made with the semi-dry wine, our added sugar might have given higher sweetness due to the different nature of the sugars. However, the differences were still perceived as small across the three samples.

This experiment is one of those cases where no clear result, or an ambiguous result, might still tell us something of value. After this experience, the very clear statements from literature and expert advice about using dry wines for cooking are at least not strengthened. But the experiment has spurred us to consider more of the wine's characteristics than only its sugar level before choosing. After all, wine also contributes with acidity, bitterness and, in the case of red wines, tannins/astringency. In her PhD thesis, Guro Rongså also states that the type of food plays a key role when selecting the wine type. It would be strong to say that we have falsified the claim of using dry wines for cooking and that it should thus be deported to the land of myths. So, if we want to produce concrete advice for choosing white wine for cooking, further experimentation is clearly required. Anyone interested in embarking on a doctoral degree on white wine in cooking, perhaps?

Basic risotto

Ingredients

½ dl olive oil
75 g shallots, chopped
300 g Arborio/risotto rice
2 cloves garlic, crushed
2 dl white wine
Ca. 8 dl hot chicken or veal stock
Salt and black pepper
60 g grated Parmesan cheese
30 g butter

Procedure

Fry the shallots gently in the oil until soft but not brown. Add the rice and continue on medium heat for about 3 minutes while stirring. Add the garlic and let fry for a short while, without it turning brown. Add the wine and stir while heating until the wine is absorbed by the rice. Add the hot stock little by little while constantly stirring, allowing the rice to absorb the liquid before adding more, until the rice has reached the desired texture (al dente). This might take ca. 20 minutes. If you run out of stock before the rice has reached the desired texture, continue by adding boiling water and stirring the same way.

Remove the pan from the heat, mix in the butter and parmesan cheese. Add salt and pepper to taste.

14

Taking Stock of Broth

Most self-respecting chefs would say that the only proper thing to do is to make the stock yourself. Or at least few would admit taking the shortcut of popping in a couple of bouillon cubes in their sauce or stew ("broth" and "bouillon" are used more or less synonymously even though "bouillon" is also used for the dehydrated cubes and powders. "Stock" is used for a more intense version where broth/bouillon has been simmered down. Broth and liquid bouillon can be used as dish in itself, while stock is considered an ingredient). Stock is made by quite humble and inexpensive ingredients, and the procedure is quite straightforward, dissolving flavor from meat, bones, vegetables, and herbs into hot water. Or is it really? Professor Hervé This, who has spent years collecting thousands of claims about cooking, found that cooking stock is among those processes that have the highest number of claims and rules connected to it in the French tradition. Stock is the cornerstone of French cuisine, the foundation that the rest of the course is built upon. While you make your stock, you will meet several points calling for you to make one or more choices: should you put the meat and bones in cold or hot water? Is the lid to be on or off the pot? Should you add the spices now or later? And so it goes.

Although the classic way of cooking a brown meat stock makes use of modest ingredients, it takes a certain amount of labor and extended time. Following a five-step procedure, cooking a stock of "restaurant

Main steps in common preparation of reduced brown meat stock

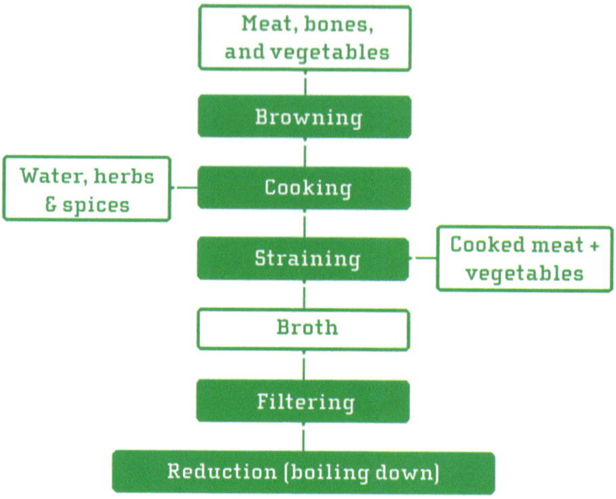

^ Main steps in making stock, adapted from Snitkjær (2010).

quality" may take a couple of days, often having the process going on overnight. The meat bones for such a stock would normally be beef or veal and classic vegetables are onions, carrots, and celery, called *mirepoix* in French. The herbs are usually parsley, bay leaves, thyme, a *bouquet garni*. You brown the meat bones and vegetables in a pan or oven to achieve the color and complex flavors from Maillard and other reactions. You then add water, herbs, and spices and let it all simmer for a certain amount of time while skimming off proteins and other components rising to the surface as a foam, and then strain it to get a rather dilute and bland-tasting broth. You then let it simmer or boil while skimming until it has reduced to a certain volume and developed a rich flavor and golden-brown color (see figure).

Stock making is not very different today than as it has been described in cookbooks throughout the last several hundred years. Why not? Perhaps because the old methods do indeed produce a very good result and thus there have not been a pressing need to change the procedure? Or maybe out of respect, or fear, of the old culinary masters? On the other hand, this way of thinking may also prevent possible progress. What if we had used this approach to the introduction of the printing press, the television, the personal computer, or the motorized car? All were met with claims that they were not necessary, or even good, for humans. Hasn't science done anything to help us learn more about the time-consuming process of making stock? Science and technology have given us the bouillon cube but have apparently not produced much new knowledge or technologies that has reached home - or restaurant kitchens for making stock from scratch. Hervé This has stated that food science has traditionally been mostly concerned with the agricultural and industrial sides of food and not found the activities of the kitchen, be it in a restaurant or at home, worthy of scientific study. That is, until the 1980s and the introduction of the branch of food science termed molecular gastronomy. Perhaps the apparent lack of interest in cooking practices, from the side of food science, is an explanation for why stock making hasn't changed much while the societies around it have changed radically?

If the chef was to embark upon comparison experiments along the five-step process of stock making with 8–10 different ingredients, she or he would end up with a vast number of possible parallel stocks to be made and compared. You may vary the type of meat, the amounts and types of vegetables, herbs, and spices, the point of addition of the vegetables and herbs/spices, browning or not browning the ingredients, starting with hot or cold water, the cooking time, cooking temperature, leaving the lid on or off, the reduction time, and probably more. Testing all these would indeed result in around 8000 parallel stocks to be cooked! For one single cook or team to carry out all these parallels is probably the project of a lifetime unless s/he had an automated research lab at hand. At least, it would probably be impossible to accomplish such a task manually and, at the same time, run a restaurant. Surely, many variations have already been tested since we today have some classical recommended procedures, vegetables, herbs, and spices to go into a stock. After all, many chefs have most likely tested and rejected many versions to find them less important or palatable. And the rejected versions would usually not end up in cookbooks.

Thanks to the recent field of molecular gastronomy and The University of Copenhagen finding this field worthwhile as investing in, the food scientist Pia Snitkjær Bailey could spend some years work-

Stock cooking commented

Finding/recommendation		Comment
According to classical procedure in cookbooks.	Meat, bones, and vegetables → Browning	
	Water, herbs & spices → Cooking	
1. Cold start is recommended: meat bones and meat placed in cold water before heating.		1. A common topic for discussion. Snitkjær Bailey found that cold start results in a clearer stock as proteins coagulate into larger particles, thus avoiding suspension of microscopic particles to give hazy stock.
2. Leave the lid off.		2. Lid kept off gives a colder surface making it easier to skim.
3. Maintain a low simmer, ideally around 85 °C.		3. 85 °C gives the most intense flavor, while higher temperature (95 °C) produces off-flavors.
4. Simmering for one hour is enough.		4. This is a radical finding! It is common to let bones and meat simmer up to 24 hours. Unless meat pieces were large, increasing the time for this step did not give any increased flavor.
5. Do not add salt yet, postpone to after the reduction step.		5. Adding salt too early may give reduced solubility of flavor compounds, resulting in them coming out with the proteins that are skimmed off.
	Straining → Cooked meat + vegetables Broth Filtering	
According to classical procedure in cookbooks.		Lab methods such as vacuum/suction filtration can give very efficient and quick filtration. Some restaurants have started making use of such equipment.

Reduction (boiling down)

1. Slow reduction at mild simmer gives more intense flavor than quick reduction. It is the reduction *time* that matters, not the volume. The most complex flavor was achieved using a reduction time of 16–20 hours.	1. Recipes often advise that the stock should be reduced to a certain volume, e.g. to half volume. This may be done quickly (rolling boil) or slowly at a slow simmer. The volume reduction is thus not the best measure. Time saved in the first simmering step can be used here to promote flavor complexity.
2. Long reduction time results in different types of stock becoming more similar.	2. During reduction, primary aromas evaporate and secondary flavors and aromas common among stocks develop. The tradeoff is thus that the shorter reduction step maintains the original character of the ingredients while longer reduction contributes to higher complexity.

^ The figure shows the steps of cooking stock together with recommendations and comments for each step. Based on the studies by Snitkjær Bailey (2010).

ing with meat stock. Indeed, she wrote her whole doctorate on the topic and revealed some surprising findings. Her thesis was divided into two parts: (1) the process and steps in cooking meat stock; (2) the effect of red wine in cooking meat stock. In her research, she obviously had to delimit her studies to focus on a few parameters to finish her thesis in time. But considering the interesting results that came out of it, her choices seem to have been very good. The findings that seem to be most surprising to experienced chefs are the impact of cooking times for the two boiling steps. Her results may indeed have a major impact on both time consumption and organization of this important activity in a restaurant. And it is definitely useful if you want to make your own stock at home. A summary of some of her findings with focus on procedure is given in the table above.

It is an interesting reflection for us that, just as in the case of cooking eggs in water, this research had not been done earlier. After all, it is about a practice that is carried out in thousands of restaurants and homes across the world every day, and the research provides knowledge that may change those practices in fundamental ways. If only these stock-makers had all been aware of it. Surely quality science communication may be an important activity.

Making a red wine sauce. There are at least two basic ways to make a red wine stock.

1. You reduce a mixture of wine and chopped onions. Add meat stock prepared beforehand and continue reducing until you are satisfied.
2. Add red wine to your broth after the straining step and let them reduce together from the start.

If you are running a restaurant, the first approach may be convenient since you can have a basic meat stock that is used in various ways - with or without wine. If you make stock for a single meal at home, you could choose the one you find most convenient.

Pia Snitkjær Bailey wanted to examine how red wines with different flavor characteristics affected the flavor of the stock. And since tannins react with proteins upon heating, she was also interested in studying how red wines with different tannin levels affected the meat proteins. She analyzed ten different stock samples based on beef stock combined with two different wines chosen for their different tannin levels: a Cabernet Sauvignon (high in tannins) and a Zinfandel (low in tannins).

She found that the aroma of the red wine was to a large degree lost in the reduction step. This is quite self-evident because volatile substances, which are those reaching our nose, evaporate easily. After all, if they didn't evaporate we wouldn't be able to smell them. The advice is, therefore, that wine for cooking should not be chosen based on its aroma, but rather on its mouthfeel and taste, such as sweetness, acidity, and astringency (the drying sensation in the mouth imparted by tannins). So, if you want to make an effort choosing a wine for your sauce, hold your nose while evaluating it. This would probably require that you are an experienced taster and sauce maker, however. Consequently, the most expensive wines are most likely wasted in a sauce.

Tannins from red wine and proteins from meat stock react to precipitate, complex, insoluble substances. If the meat stock and wine

The components of a red wine stock

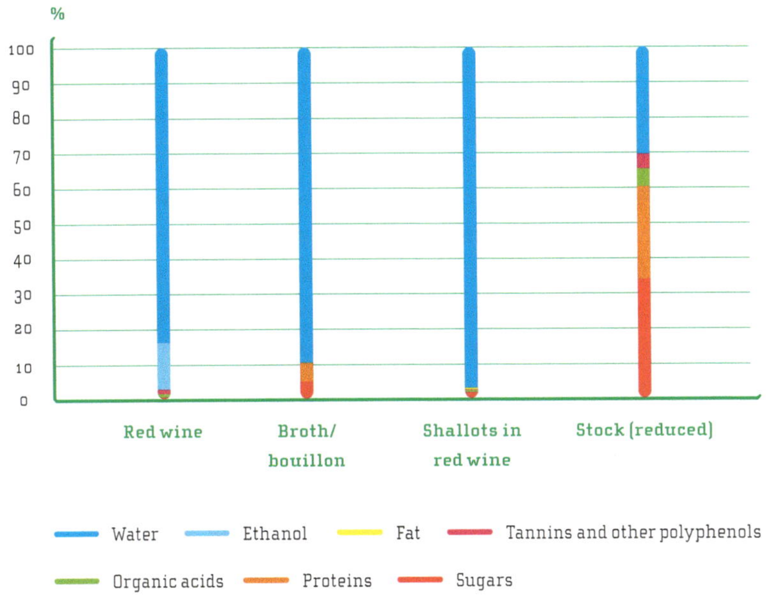

^ Before reduction, red wine and broth is mostly water. After several hours' reduction, the water content is reduced from the original 90% to 25%. At the same time, the concentration of non-volatile compounds such as sugars, acids, and amino acids increases markedly. In addition, numerous new substances are produced as result of chemical reactions.

are reduced separately and thereafter combined, the result might therefore be a slightly grainy/sandy mouthfeel. Snitkjær Bailey also reported that this procedure led to a more astringent and perhaps bitter sauce. Therefore, reducing the meat stock and wine together gives a less astringent sauce than that made by reducing stock and wine separately. Not surprisingly, this effect is more pronounced when using a wine rich in tannins. If you still choose to reduce stock and wine separately, you may combine them and reduce for 30–45 minutes and finish it off by a filtration to avoid the risk of a slight grainy and astringent mouthfeel.

A fine wine experiment Red wine sauce, similar to, for example, puff pastry and some other products, is a laborious ingredient where the industrially prepared alternatives may be competitive for the occasions where you don't have time to carry out all the time-consuming steps. Although maybe a disappearing skill in home kitchens, we were inspired by Snitkjær Bailey's work and decided to have a closer look at what happens in the pan during the reduction step in one of our food workshops. We were curious to see with our own eyes, and taste with our own palates, which roles the red wine could play in the sauce.

The point of departure was a basic recipe for red wine sauce: take half a liter of good meat stock and half a bottle of red wine and reduce it down to one deciliter. The procedure is not ideal according to Snitkjær Bailey's research, but the approach would give us two stocks that would be directly comparable in terms of reduction time. To which degree would the characteristics of the wine affect the flavor of the final product? Would minor differences in acidity, sweetness and mouthfeel in the wines show up in stocks cooked for several hours? If so, how are the differences experienced?

We prepared three different red wine stocks with three different red wines with increasing body, aroma intensity, and amount of dry extract: a light Beaujolais, a medium-bodied Rioja, and a full-bodied Argentinian Malbec. The differences in acidity, sweetness, and body were clear but not dramatic. For example, the acidity varied from moderate (5.0 g/l total acids in the Beaujolais) to high (5.7 g/l total acids in the Malbec). The dry extract is the amount of solids left when all water is evaporated and gives indications towards the quality and body of the wine. In these wines, this parameter varied between 24 g/l in the light-bodied Beaujolais to 35 g/l in the full-bodied Malbec. The Rioja was in between these two extremes both in acidity and dry extract content. Would these differences reflect in the stock when tasted by the panel consisting of workshop participants?

The sauces were cooked, coded, and served as blind samples for the tasting panel. The differences in sauces were obvious already to the naked eye. The Beaujolais came out lightest, the Malbec darkest. The participants were asked to rank the three sauces in order of increased body, bitterness, acidity, sweetness, and aroma intensity. The values were summarized and displayed as a radar diagram often used in sensory science to visualize how the different characteristics were evaluated by the panelists.

Blind tasting of three red wine stocks

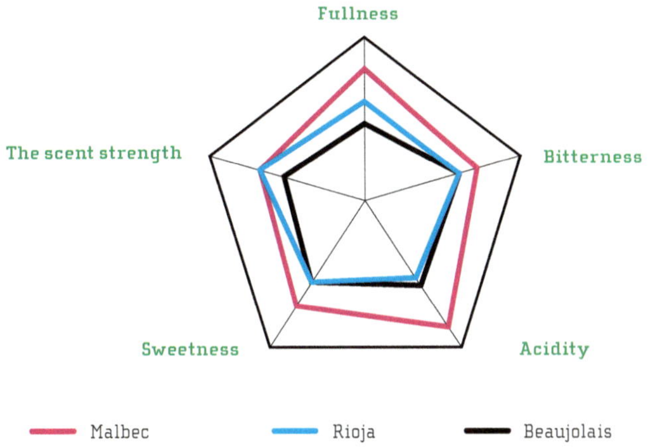

^ Results of red wine stock blind tasting (panel of 12 participants): The radar diagram shows that the three wines gave stocks with very different sensory characteristics.

The differences in the three red wine sauces were quite obvious, and there was no question about the important role of wine in cooking (see diagram). The full-bodied Malbec gave, except for aroma intensity, a stock with more of everything: body, acidity, sweetness, and bitterness. The Rioja gave a stock that was fairly similar to the one with the lightest of the wines, but somewhat more body and clearly higher aroma intensity. The result was quite unambiguous: the wine makes the sauce!

Naturally, we also asked the question that most choose to ask right away: which stock would you prefer? Liking is always a matter of personal taste and more or less up for debate. In this case, nine out of 12 participants considered the stock made with the full-bodied Malbec as the best, the three remaining preferring the Rioja-based stock. However, this time, the liking-issue raised some harmonious debate among the participants. Everyone agreed that a sauce should match the main ingredient of the dish and that it cannot be evaluated alone. Every dish requires its own tailored sauce.

15

Salt Shapes the Pasta

How does a non-Italian know when the pasta is ready? S/he throws it on the wall and if the pasta sticks to the wall, it is ready. More often than not, jokes contain a kernel of truth. This one describes two important things about pasta: 1) Italians love pasta and have a passionate relation to how to cook it properly; 2) pasta is considered overcooked when it becomes gluey and sticks to the plate, the wall, ceiling, and so forth.

Italian recipes for cooking pasta usually put a great emphasis on achieving the ideal texture. The exactly right *al dente* structure, literally "for the teeth," means that the pasta should have a certain firmness or resistance to the bite, but without being hard or brittle. In the case of dry pasta, the *al dente* structure is said to be reached the moment that the hard and white center in pasta is no longer visible. Careful monitoring and testing of the softness of pasta using one's teeth during cooking is a common advice to ensure a good structure. Another advice is to stop the cooking slightly before the pasta reaches the desired structure, as the cooking continues for a short while after the pasta has been strained. For an Italian, cooking pasta can be a play with seconds, whereas many others would only count minutes. No wonder they make jokes about the pasta-ignorant people from other countries.

Pasta = starch trapped in a network of proteins. The most important chemical substances in pasta are water, starch, and gluten proteins—the very same elements that make up a good bread. Both pasta and bread are built up by a strong and elastic gluten network. Microscopic starch granules in the flour swell as they absorb water during heating and get trapped in the gluten network.

Overcooked pasta becomes sticky because the starch granules absorb so much water that they burst open, leaking out long starch molecules that form an adhesive, sticky layer that makes individual pasta strands or pieces stick to each other, and, for that sake, to the wall. In a properly cooked pasta, the starch granules are swollen and soft, but still intact, with the starch molecules in place inside the swollen granules. Gluten does a good job of keeping the starch in place, inside the pasta, rather than letting it escape and float around in the water as a thickening agent.

Gluten is a strong network of wheat proteins permeating the entire dough to make it both plastic and elastic. This network gives a dough that can change shape upon pressure or stretching (plastic), but can still return to its original shape when the pressure of stretching is relieved (elastic). To build such a network, we need a good amount of gluten-forming proteins. Untouched, these proteins are globular in shape, like a ball of woolen thread twisted and folded and swimming in the wet dough slurry of flour and water. In a watery environment, part of these proteins (glutenins) unfold and and entangle to form a web-like network that ultimately permeates the entire dough. The web-forming reactions speed up during kneading, surely noticed by any baker. The second type of proteins (gliadins) keep their original ball-like shape but bind lightly to the glutenin network. Solubilized in water, these small balls increase the viscosity of the dough, and they act like plasticizers, tiny ball bearings, in the network, thus enhancing both the plastic and elastic nature of the dough. A dough with a strong gluten network will stretch when you pull it but return to its original shape once you release it. Just imagine the wonderful dough in the hands of an experienced pizza maker, at the same time malleable like a play dough and elastic like a rubber band. The gluten network is this rubber band; actually, it is thousands and thousands of microscopic rubber bands bound to each other as a three-dimensional network throughout the dough.

Gelatinization of wheat starch in pasta

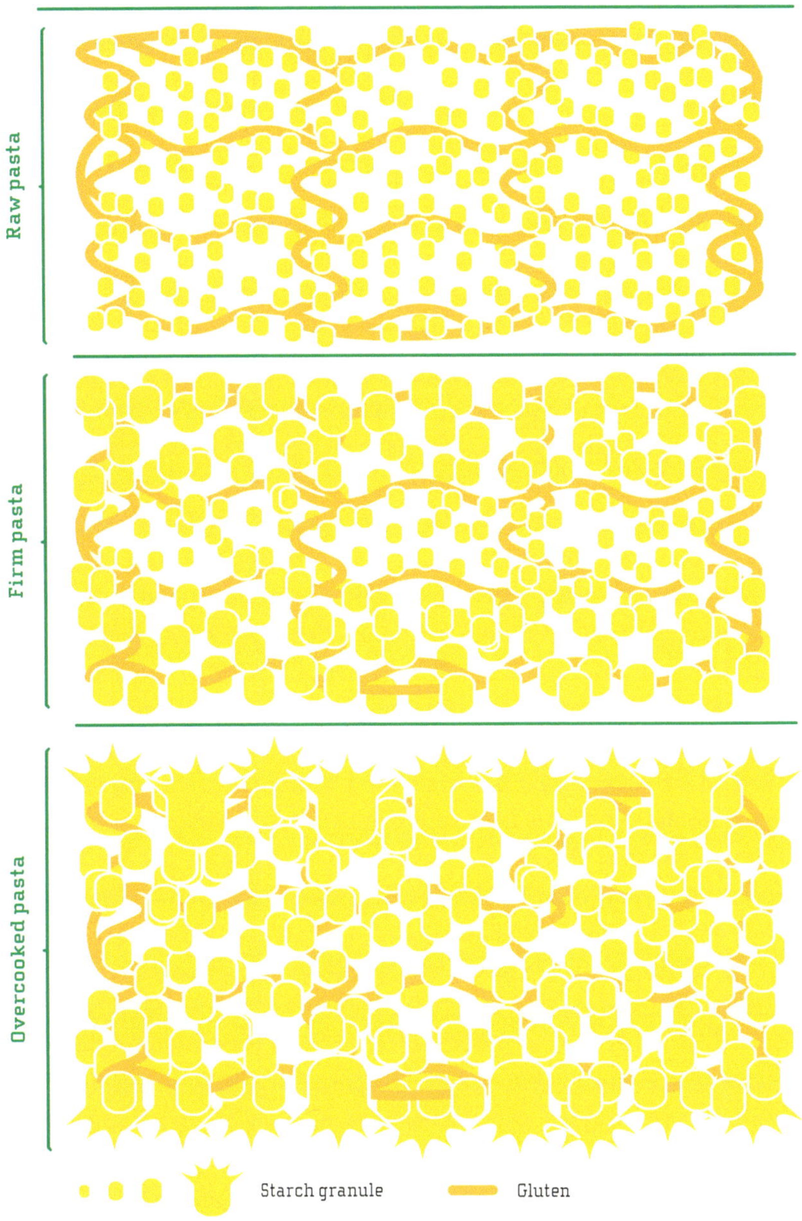

^ Starch granules in pasta swell in hot water to give a gel, kept in place by the gluten network. The chef's challenge is to stop the cooking at the moment where the granules in the middle have swollen just enough, while those at the surface are still intact. Otherwise, the pasta will be too hard in the center (undercooked) or starch molecules leak out at the surface to form a gluey layer.

What about salt in the cooking water? The role of salt seems to be important in Italian pasta recipes especially, as they seem to contain significantly more salt as compared to, for example, most Nordic pasta cooking recipes. The Italian rule of thumb is 100:10:1, where the numbers indicate the relative amount of ingredients: 100 grams pasta requires 10 grams salt in 1 liter of water. In a world where we are commonly advised to cut our salt intake, this sounds like an awful lot. Indeed, many recipes seem to manage with significantly less salt, and ratios as low as 5 grams of salt for 2 liters of water are not uncommon. Compared to the Italian rule of thumb, 100:10:1, we have seen recipes asking for as little as 100:2.5:1.

Why are these recommendations so different? Looking through literature and various cooking forums revealed different sorts of reasoning behind the role of salt in the pasta cooking water. A common reason is obviously taste, and salt is claimed to be absorbed by the pasta. This sounds logical as dry pasta absorbs significant amounts of cooking liquid, and that other flavorings should tag along with the water into the pasta. Another explanation is related to the structure. Salt is said to help pasta remain firm and less sticky during cooking. This is usually explained by the fact that salt is known to retard the gelatinization of starch, the phenomenon where microscopic starch granules from the flour absorb water. Thus, pasta cooked in salty water does not so easily become sticky. Third, salt is known to promote formation of the gluten network. Salt ions interact with the charged parts of the glutenin molecules, to decrease the repulsive forces between glutenin molecules. This allows them to come closer to each other and thus react more easily to give a strong gluten network.

In summary, salt might affect the structure in pasta either to strengthen the gluten network, to prevent the starch becoming gluey, or both. However, in many discussion forums, there are many claims about salt having no effect on the structure of pasta whatsoever, and we would assume that at least some of these should be based on experience. It seems that we have a nice mixture of contradictory claims, which clearly asks for an experiment, and a suitable topic for a food workshop. We decided to let the following "double research question" guide our inquiry: does salt have a firming effect on pasta structure as claimed or is the role of salt, and thus difference in recipes, only a matter of taste?

The pasta experiment. We invited participants to a workshop and cooked pasta in four pans with two different concentrations of salt. Two pans had salt content based on the Italian recommendation 10g/l and the other two contained cooking water with a lower salt content, namely 2.5 g/l. In order to expand the experiment we used two different types of pasta: one made of durum wheat traditionally used for pasta and another made of regular wheat. These two wheat varieties should expectedly give different structures to the cooked pasta. While normal wheat flour contains approximately 12% protein, durum wheat has a protein content of around 15%. Also, the starch is slightly different between these two types of wheat. The starch in durum wheat has a lower gelatinization temperature, and it also binds more water than the starch in common wheat. So, this could potentially give us two answers from one experiment: we might get a clue of both the effect of salt and the effect of wheat variety on the structure of cooked pasta. The four parallels were:

- Regular wheat pasta cooked in low salt (2.5 g/l) water for 3 minutes.
- Durum wheat pasta cooked in low salt (2.5 g/l) water for 3 minutes.
- Regular wheat pasta cooked in high salt (10 g/l) water for 3 minutes.
- Durum wheat pasta cooked in high salt (10 g/l) water for 3 minutes.

The pastas were freshly made at a small pasta factory in Helsinki based on a recipe of 70% flour, 20% eggs and 10% water. The pasta dough did not contain any salt to ensure that the only salt experienced by the pasta should come from the cooking water.

The four parallels were coded for blind tasting and served with good tomato sauce made by chef Tatu. The 16 participants of the experiment were asked to evaluate which parallels were the softest and the firmest of the four alternatives. We also asked them to state which they would prefer most and shortly explain their choice.

Indeed, the results convinced us that both the salt level in the cooking water and the wheat variety used are important parameters in the quest for a firm structure in the ready-cooked pasta. Ten out of the 16 participants on this evening evaluated the durum pasta cooked in high-salt water to be the firmest, and nine considered the regular wheat pasta cooked in low-salt water to be softest. The pasta made of regular wheat flour cooked in high-salt water also got a few votes as the firmest, indicating the salt contributes to a firmer pasta even

Blind tasting of pasta

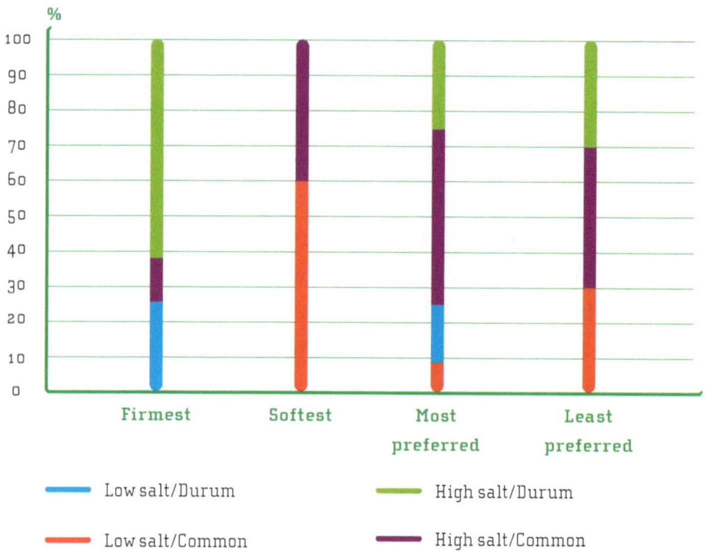

^ The illustration summarizes the result from the pasta experiment, percent of votes. The two leftmost columns/bars show that both type of flour and amount of salt affect the firmness of the pasta, but that type of flour most likely has a greater effect of the two. The two rightmost columns show that personal taste preferences vary considerably.

if the flour is not high in protein. The important role of high protein flour became obvious as neither of the durum pastas got any "softest pasta" votes.

Within pair comparisons of wheat variety and salt level also gave an indication of the importance of these two parameters. Of the two durum wheat pastas, the high-salt version received more votes as the firmest (green vs. blue bar), and within the regular wheat versions only the high-salt version received votes as firmest (purple vs. red bar). When comparing the two durum wheat pastas on the one side (green and blue) with the regular wheat versions (purple and red) on the other, the former dominate the votes for firmest, while the latter are the only ones receiving votes for being softest, clearly showing that the wheat variety is important. But which parameter is the most important; the amount of salt or wheat variety? If we compare the durum wheat pasta in low salt (blue) with regular wheat pasta in high salt (purple), we get some sort of measure of this. Both received votes for being firmest, but of the two, only the high-salt regular pasta received votes for being the softest of all. With great caution, we could say that the nature of the wheat has a somewhat stronger effect on the texture than the amount of salt in the cooking water. The salt content obviously affected the taste of pasta, so the high-salt versions were easy to identify. From an experimental perspective, this may be source of error because the tasting was not entirely blind. Some panelists might have been affected by their preconceptions, conscious or unconscious. The evaluation of wheat variety was hence more unbiased because the two high-salt and the two low-salt parallels were compared internally.

The liking votes were, as usual, rather mixed, and an interesting result was that the high-salt pastas received approximately the same number of votes for most and least preferred. The high-salt regular wheat pasta appeared as particularly controversial in this respect. Yet again, we learnt that even though analytic tasting can give pretty clear results, we part when it comes to personal preference.

Which recommendations could we give to those appreciating a good, al dente structure in their pasta, while at the same time maintaining a low salt intake? It seems that the Italians do have good reasons to recommend a relatively high amount of salt in the water. But it helps if you use durum wheat pasta high in gluten-forming proteins, and make sure to limit the cooking time, as using low-salt water might more easily give overcooked pasta. Even if you do, an Italian still might choose to throw the pasta at the kitchen wall.

Can we avoid sticky pasta if we add oil into the pasta cooking water? It is often suggested that oil in the cooking water would give an oily surface to the pasta which would then prevent the individual pasta strands or pieces from sticking to each other. However, the oil floats on the surface of the water and cannot reach the pasta immersed in the water. Oil will help against sticking if added after draining off the water. The oil will also contribute to depth and richness of flavor and a smooth texture. If you want to avoid sticky pasta, use copious amounts of water, enough salt, and stop the cooking in time. ×

16

You Can See It, Hear It, Taste It and Feel It— What Is It? It Is the Taste of Food!

Let's begin with a story. Once, not very long ago, in a restaurant, a sommelier poured Anu two different red wines and asked her to tell which wine went better with the main course. Already, as she watched the wine gushing from the bottle into the glass, Anu knew which one she would prefer. The decision, made unhesitatingly in a split second, proved right. The wine she had chosen went beautifully with the dish while the other wine tuned both the wine and the dish down to somewhere between ordinary and bland.

What did the chemist base her decision on? The color of the wine, the viscosity, the shape of the glass or the sound of the wine being poured? Did her eyes, ears or nose register something that her analytic self did not have time to parse or analyze? Perhaps the decision involved all of those sensory modalities at one and the same time. What she knew for sure was that the unhesitating decision was made by someone else than her conscious self. Her rational half had willingly taken the role as passenger in this multisensory trip where someone else was doing the driving. That "someone" guided her through the stream of senses where she could not tell where one perception started and another faded away. But whether the driver was her mythical subconscious self or some yet unexplored neurochemical phenomena, the trip was nevertheless an enjoyable flow of senses. To everyone who claims to be a rational and analytical person we recommend these types of moments, to leave the intellectual mind behind and let the sensations and emotions be the guide. A welcome change from the routines of everyday life!

Food, and why not all other pleasures as well, are produced by the joint effort of several of the senses. To "taste" the food when experiencing our meal, dish, or snack, we actually use all our five senses - sight, smell, hearing, taste, and touch. These sensations do not operate in isolation but intertwine so that one sensation influences the quality and intensity of the other. Furthermore, these sensations are connected to our memories and values, which further influences how we "taste" our food. Losing only one of these will fundamentally alter our experience. You can experience the importance of smell in food if you have a cold, or by pinching your nose, while chewing the food. Using only our sense of taste in the mouth, it can sometimes be difficult to identify which food we are tasting. Is this apple or grape juice? When we then recover from the cold, or stop pinching our nose, the smell sensation returns and the food becomes identifiable. In other words, smell and taste together create an experience that we, in layman's terms, call food's "taste." Vanilla has virtually no taste but an intense aroma. So, when you're enjoying a vanilla ice cream, it is not a vanilla-tasting ice cream; you are having a vanilla-scented ice cream. Likewise, the experience of lemon consists of both taste (mostly sour) and its characteristic scent. Taste and smell sensations also influence one another so that by combining a specific odor with food we can enhance or suppress tastes such as sweetness or saltiness. For example, smells that we have accustomed to associate with sweetness, such as strawberry or banana, can enhance sweetness in food. Similarly, odors commonly experienced together with saltiness can enhance the sensation of saltiness, and even induce the notion of a salt taste where there is no salt (more about this in the chapter on reducing the amount of salt in your food).

According to current knowledge, we have taste receptors for only five taste characteristics, namely sweetness, sourness, bitterness, saltiness, and the savory umami taste (there is recent research that indicates that we may have receptors for other experiences as well, such as fat and metallic taste). Even though we have a limited number of tastes, we are able to discern, for instance, between different types of sweetness, such as the difference between sugar and aspartame, stevia or fructose sugar. So perhaps we need to start talking about different shades of sweet, sour, bitter and so forth. Smell is even more versatile, as the most recent research suggests that we can identify tens of millions of different smells. We have a limited number of basic taste sensations, but we don't have "basic smells." When it comes to the nose, the closest we get is to go by association, be it particular things or experiences, such as "lemony aroma" or "wet dog." This is the reason that those describing aromatic foods and drinks such as wine and coffee use such flowery language when describing a product: "This Ethiopian coffee has character of bergamot, citrus, and jasmine, with a distinct lingering sweetness and rounded acidity." Our brains also receive other signals such as the temperature and texture of food that we feel through the trigeminal nerve system while chewing our food. The chemical burning sensation of, for example, chili interacts with our heat sensing nerves: "This is hot!" And this is also embedded in the language as we use the same word for the two. The cooling sensation from menthol comes from the chemical interaction between menthol molecules and our temperature receptors ("menthol" is an ambiguous word describing a concept as well as a distinct chemical compound found in various plants). The tingling sensation from carbonated drinks appears due to mild stimulation of the pain receptors in our mouth. All these work together in an interplay to give us the experience that we call the *flavor* of food. ×

Senses of man, unite! In addition to smell-taste pairs, sight perceptions have been shown to blend together with taste into a merged (multi)sensory experience. For example, the red color in food has been observed to enhance sweetness whereas green enhances the intensity of sourness. Don't be disturbed, though, if your personal feeling is that green is bitter and red is sour, as these "tastes of colors" are very strongly based on learned associations that we start making from the very beginning of our life. There is no right or wrong when it comes to personal sensory associations. However, when many people are asked, common patterns do emerge. These learned associations also have an effect on how we taste our food. Thus, if strawberry jam is the red food that you usually have, you will most likely experience the color red as sweet, and red color in your food may even enhance the intensity of the sweetness in your food. However, if your experience of red fruits is mostly lingonberries or cranberries, you might experience red as sour or bitter. There are a number of studies published to show such mechanisms. For example, how a red drink has been shown to be experienced as sweeter than a yellow or colorless one, even though the ingredients and thus its actual sweetness were identical.

Vision may lead our perceptions in other ways. The French researcher Gil Morot and his coworkers asked oenology (wine studies) students to describe aromas in white and red wines. When describing the white wine, the students used terms such as honey, lemon, grapefruit, straw, banana, sulphur, butter, and mango. The red wine, on the other hand, was described using words such as coal, blueberry, plum, cherry, Havana cigar, chocolate, cocoa, and violet flowers. The interesting part of the study was that both wines were actually the same white wine, and if you had tasted them with your eyes shut there would have been no detectable difference. The apparent red wine had just been colored red with purified grape anthocyanins. An elegant experiment that shows how powerful the sense of vision is in guiding our other senses.

Most of us would probably accept the fact that the smell or even our sight intertwines with and influences how we taste our food. However, it might be harder to believe that our hearing also has a similar effect. The sound of pouring the drink into the glass, the fizzing sound when you open your soda or the crunching sound of the carrot influences the experience of taste, liking or freshness of our foods and drinks. Researchers have demonstrated the independent role of sound by studies where only the sound of the food or drink is changed while all the other characteristics have be kept constant. The Danish researcher Kristian Holt-Hansen described, as far back as in the 1960s, how the pitch of a beer's fizz affected the experience of beer. The study first showed how two different Carlsberg beers, Lager and Elephant, both have their unique fizzing sounds although beer drinker most likely don't pay attention to this fact. However, when the fizzing sounds of the two beers were switched, the taste panel rated both beers as significantly less pleasant. Obviously, although the participants did not recognize that the sound had changed, they must have noticed that something had changed in their sensory trip to their pint, and thus they reported decreased pleasantness over the same foamy gulp but with the wrong fizz. This is a challenge for you dedicated beer drinkers: can you differentiate the sound of your favorite beer from other brands?

Another classic study is Massimiliano Zampini and Charles Spence's study published in 2004 on how both the sound volume and frequency of the crunchy sound affected respondents' experience of crunchiness and appeal when chewing potato crisps. The participants sat in a cubicle enjoying potato crisps while wearing headsets, and were asked to evaluate the quality and liking of the different potato crisps. The sounds from the respondents' mouths were recorded by a microphone and played back simultaneously through a headset. This way the researchers could turn up and down the volume of the crunching sound, and they could also selectively amplify or reduce specific frequencies. Louder crunching sounds, as well as selectively amplifying high-pitch frequencies, resulted in the participants reporting the crisps to be crisper and fresher. Zampini and Spence's study might make you laugh, and it did since it was awarded the Ig Nobel Prize in nutrition in 2008, a series of prizes awarded "to honor achievements that first make people laugh, and then make them think." But the question behind it is also serious: when one sense affects another, is it to be considered a false sensation or a brand new complex sensory perception? For example, does the red liquid *actually* taste sweeter than the green with the exact same amount of sugar, or is it an illusion? Or does the soup with sardine flavor really taste saltier, or is your sensation false? The basis for all of this is that the sensation does not occur only in our mouth or nose, but in our brains where all the different sensory modalities—sight, smell, taste, touch, and hearing—unite into one unique multimodal sensation.

Wine spiced with music or music seasoned with wine? The lucky participants of another study by Spence and coworkers were served some excellent wines while listening to a concert by the London Symphony Orchestra. While enjoying both music and wines, the participants were asked to tell which of the musical pieces and wines matched. The participants reported surprisingly similar pairs of music and wine. This made the researchers continue with more experiments by selecting the most frequently matched pairs of wines and music and then asking participants to evaluate the wine with and without music. The participants were asked to evaluate the wines in terms of typical characteristics such as sweetness, acidity, fruitiness as well as their own enjoyment of the wines. The results indicated that the participants perceived the wines sweeter and more enjoyable while listening to the matching music than while enjoying the wines in silence. Indeed, one could imagine that music matching emotionally for champagne would be different from the best musical match for red wine. This is again a great result to build upon, which we did in our food workshop.

Since our senses apparently work together in both ordinary and extraordinary occasions, we asked ourselves, inspired by Spence and coworkers' studies: "is a red wine enjoyed alongside one of Mozart's piano concertos the same red wine as the same bottle enjoyed with one of Tchaikovsky's string quartets?" As we did not have a symphony orchestra to work with, we did "a poor man's version" by selected recordings of the same music: Mozart's flute quartet in D major KV285 second movement (allegro), and Tchaikovsky's string quartet No. 1 in D Major, second movement (andante cantabile). The Mozart movement is an up-tempo piece in a major key for flute and strings, giving a light and cheerful/merry character. It's a very typical Viennese classic: gracious, elegant, and orderly/neat—giving associations to pastoral and aristocratic/noble scenarios. The Tchaikovsky piece for four strings is in a slower, walking tempo in major key. Compared with the Mozart piece, it is more toned down, at times melancholic and nostalgic, but with some more rhythmic sequences, like a beautiful hymn. The music-lover might say it could connote nostalgia and romance. The wines of our choice, selected to resemble the wines in the original study within our more modest budget, were a soft, tannic, fruity and slightly spicy red Bordeaux Château de Sequin 2011, and a crispy, dry white Laroche Sauvignon Blanc de la Chevalière 2012 with marked acidity.

The tasting was conducted in three stages. First, the participants were asked to taste and evaluate the characteristics of the wines. They were asked to focus on sweetness, acidity, fruitiness, astringency (tannins), alcohol character and aroma intensity. They were asked to rate the wines according to a scale from 1 to 3, where 3 is most and 1 is least. Thereafter the participants were asked to reflect upon which type of music that would fit the wines, and which instrument the wine could represent if it was an instrument. The wines were indeed characterized by different music and instruments! The red wine was associated with progressive metal music, Bach, soft jazz, Sade, Adele with cello accompaniment, male choir, Apocalyptica and "the music police standing in a rock concert." The white wine was characterized by swing music, ABBA, Vivaldi's Spring, a picnic concerto, Chopin, Schubert's Die Forelle, or Monteverdi's Orfeo. The instruments chosen for the red wine were lute, cembalo, cello, or saxophone whereas the white wine was represented by flute, soprano saxophone, or steel-string acoustic guitar.

After the descriptive session, we turned the participants' attention to other things so that they would take some distance to the wines. After approximately 10 minutes of relaxed chatting, we refilled the glasses and played the Tchaikovsky movement, asking them to evaluate numerically the same characteristics for a second time. After this second tasting session and another 10-minute break of chatting, we asked the participants to numerically evaluate the wines for the third time now listening to Mozart.

First, it is worth noting the spontaneous comments from the participants: "How come the flavor becomes more fruity with this music?" and "The wine smelled stronger when the music changed."

Although the effects were not strong and the number of participants too small to carry out a proper statistical analysis, it still seemed as if the music had some effect on the taste characteristics

The effect of music on wine

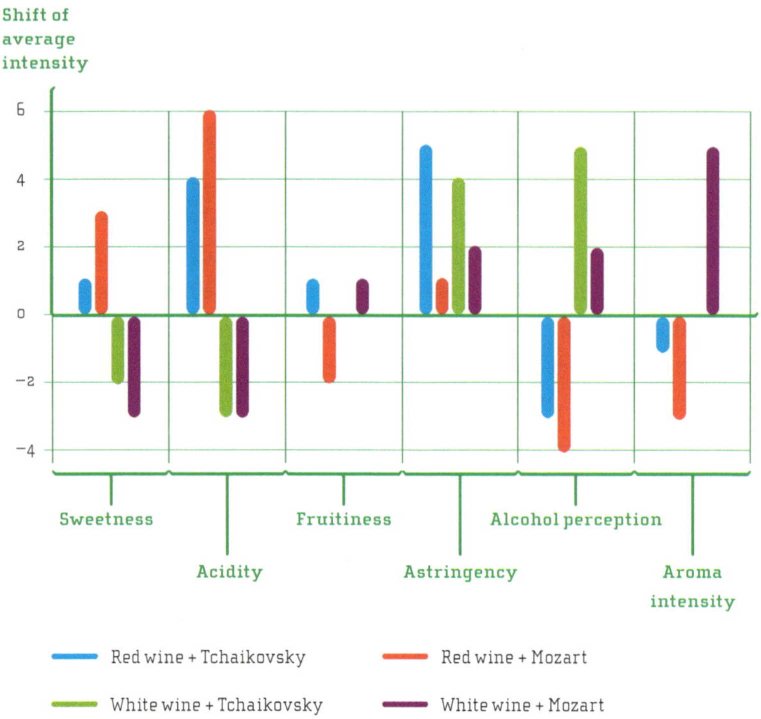

^ The diagram shows the effect of music on the characteristics of wines compared with tasting in silence. The higher the bar, the greater the effect of the music. A positive value means that the music enhances and a negative value means that the music suppresses the given characteristic. Close to zero values indicate very low or no effect of music on the sensory perception of wine.

of the wines. The bars in the diagram indicate the shift of the average intensity on the various characteristics of the wines affected by music compared with tasting without music. The higher the bar, the greater the effect of music compared with silence. A positive value means that the music enhances the given characteristic while a negative value means that the music suppresses the given characteristic. Close to zero means that the response saw, on average, little or no link between taste characteristic and sound. We considered a change to be significant if the value was higher than 2 units; if the average value changed only 1–2 units in one or another direction, then the change was considered not important. Thus, Mozart seemed to enhance the aroma intensity of the white wine aroma while Tchaikovsky enhanced the sensation of astringency in both wines. Acidity in the red wine seemed to be enhanced by both Mozart and Tchaikovsky, and perhaps suppressed in the white wine, as compared with tasting in silence.

Our 16-member panel indeed gave, in many respects, similar results to what the original study by Spence and coworkers had indicated: slow, lingering, "in front of the fireplace" –feelings were dominant for the red wine whereas the white raised light and cheerful feelings. The participants had few problems in associating musical instruments and music with the two different wines. The opinions were so uniform that one would think they were secretly communicating with each other without us noticing. The participants were also very like-minded about the music-wine pairs: 14 out of 16 matched the red wine with Tchaikovsky and the white with Mozart. No wonder that wine critics describe wines by intuitively picking expressions and metaphors from music. The same should apply to other foods and drinks as well. After all, how would you consider a steak if a poetic food critic described it with the words: "This steak tastes like driving a Ferrari with flat tires?"

17

Can You Cut Down on the Salt Without Affecting the Taste?

Consumer statistics and nutrition science tells us that we eat too much salt, 8–10 grams a day, twice the amount that we should. But what can you do? Salt tastes too good! In addition to being salty and, as such, very likeable, salt is a flavor enhancer on a deeper level than just the salty taste itself. In a research study published in *Food Technology* in 1985, Marianne Gillette described an experiment where respondents were to evaluate chickpea soups with various levels of salt. She found that the addition of only 0.3% salt, much lower than the usual content in such soups, resulted in the soup being experienced as more salty, sweeter, richer, more balanced, and with a larger overall flavor. The soup was also experienced to have less off-flavor, less metallic taste, and less bitterness. Compared to the same soup without salt, the panel even experienced the soup to have a thicker or richer texture! The first thing we can infer from Gillette's paper is that various tastes affect each other when they're experienced in the same mouthful. Based on

Taste–taste interactions

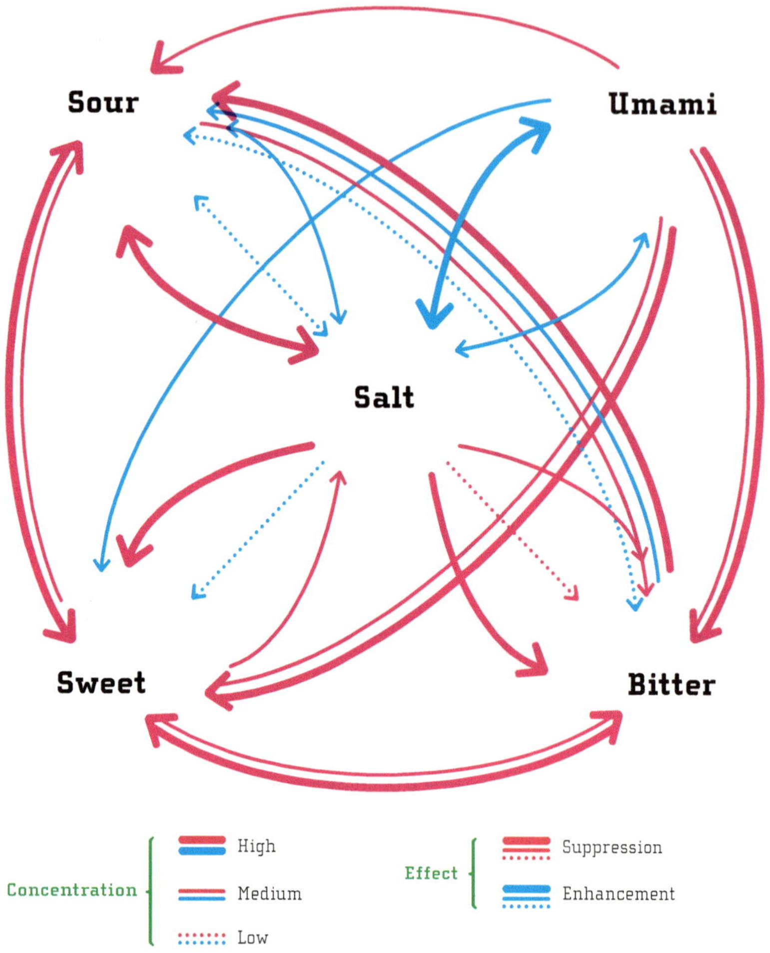

^ The tastes enhance or suppress each other depending on concentration of the tastants. For instance, the figure shows that sweet taste at low and medium concentrations suppresses bitterness—and that this effect goes both ways. The figure is based on an illustration by Martin Lersch published in his blog khymos.org.

a review article about such taste–taste interactions, Norwegian food blogger and chemist Martin Lersch made a figure that visualizes how tastes have been shown to affect each other (see figure).

These interactions are by no means simple or straightforward. As one example, salt in low concentrations may accentuate sweet taste, like when you add salt to sweet buns or the mentioned chickpea soup described by Gillette. Did your grandma advise you to add a pinch of salt in sweet dishes "to remove the watery taste?" But higher concentrations of salt can have the opposite effect and suppress the experience of sweetness. When it comes to bitterness, the sensation of salt seems to suppress the experience in both low and higher concentrations. Indeed, the word "salad" is said to have its origin from the word salt: the ancient Romans seasoned their occasionally bitter salad greens with salt, perhaps to suppress bitter taste. And salt promotes the pleasant taste of umami.

Salt seems to have some magical powers over the sensory properties of food. If we need to cut down on the salt, do we simply have to accept that our food must be less tasty, a self-imposed asceticism where flavor is sacrificed on the altar of health? Or does there exist some sort of "hedonistic solution?" Perhaps knowledge about how we experience salt can be of help to maintain the good taste and, at the same time, reduce the intake of salt?

Can we get used to less salt? Cutting down on salt is a bit like quitting smoking. Many have tried, and even succeeded, with the ascetic strategy. Others have tried and resigned, while others again don't bother trying in the first place. In a dinner party, this situation could easily result in the person who has given up, or is indifferent to her or his salt intake, having to humbly ask the host to bring the salt to the table. Or, in fear of offending the host, accept that the food is experienced as somewhat bland. This everyday experience has support in science. In a doctoral thesis from 1988, Patricia Elmer showed that preference for salt subsided slowly when the research subjects cut down on the salt. Those who cut the salt by 30%, which is a rather radical change, preferred less salt after some time. But this change was slow: after three months the respondents had changed their preferences so that their perception of the "ideal" salt level was 15% lower than before the experiment. After half a year, it had reached a 20%

reduction, and when the experiment ended after just over a year, the perceived ideal level was reduced by 25%. Conclusion: asceticism *does* work, at least to a certain extent.

Does preference for salt vary between countries and cultures? There is research to support this notion. When researchers compared the recognition threshold among people from Myanmar and South Korea, they found that there was a considerable difference between the two groups. People in Myanmar, used to eating food rich in salt, had a higher threshold (0.17%) for recognizing salt taste whereas the South Koreans, with a traditionally low-salt diet, could recognize the salt in concentrations of 0.10%, which is the commonly reported minimum threshold for what is noticeable for humans. Indeed, the researchers link this, among other things, to the presence of refrigerators in the two countries, as salt is a preservative in non-refrigerated foods. Imagine what the threshold was 150 years ago, when preserving food by salt was a common practice (practically all meat and fish was highly salted) in many places?

Furthermore, research shows that we use less total salt when we add it at the table compared to adding it during cooking. So even if the host who made the low-salt dinner put the salt shaker on the table, inviting the guests to add salt to taste, we'd probably come out with lower total salt consumption. A reason that this leads to less salt consump-

>>

On British crisps. Those who have had potato chips in the United Kingdom, the Brits say "crisps," may have noticed that some producers have printed "ready salted" on the plain salt versions. Why "ready salted" and not simply "salt" like, probably, the rest of the world would say? Yet another of those British oddities? Historically, the British crisps were packed unsalted, and you had to fish out the small packet of salt lying among the crisps and add the salt yourself before enjoying the snack. At some point, a producer got the idea that s/he could take the effort of adding the salt and use this in marketing the product. Thus, the British "ready salted" crisps were born. Do you think the crisp lovers among us would eat less salt if we were to add the salt ourselves? ×

tion might be that salt on the surface of the food is felt more intensely compared with salt dissolved in the water phase of the food. However, things are never as straightforward as they seem. Dissolved salt in, for instance, a stew will slightly change the properties of the water solution in such a way that aromas are more easily released, resulting in a dish with more intense, or less bland, flavor. For those versed in chemistry: aroma compounds are often rather nonpolar and volatile, and dissolving salt increases the water's polarity and decreases water activity. This results in the water being less able to dissolve some aroma compounds, which are then more easily released, which we consequently experience as increased flavor. Finally, salt present during cooking may very well change the course of the cooking process. In one of our porridge cooking experiments we observed that adding salt before cooking led to a porridge where the milk became slightly curdled, whereas adding salt after cooking produced a porridge with a more homogeneous liquid part. Indeed, adding the salt after cooking gave a more preferable porridge. Also, salt makes a great difference when baking bread and cooking pasta, as it contributes to reinforce the gluten network (see Chapter 15 on cooking pasta).

Reaching everybody. If the goal is to reach the whole population, the health authorities can probably not expect everybody to display the willpower that the ascetic strategy requires. Research has shown that pure health marketing reaches only a small part of consumers, whereas most of us simply turn our backs on purely health-marketed products. And if you are a restaurant chef, it is probably poor business to try to educate your guests by telling them that your food, albeit bland, is at least healthier. But perhaps we can use research-based knowledge and the taste–taste interaction diagram for our own and others' benefit. If the figure is correct and representative for everyday use, it should be possible to add a small or medium amount of something sour, which should in turn enhance the experience of saltiness in the food. But balancing the amounts carefully is important, because, as the diagram shows, a high level of sourness may have the opposite effect. A home experiment could be to make identical batches of meatballs where both have less salt than given in the recipe but then adding a small amount of lemon juice or vinegar to one batch. Will the meatballs with lemon juice or vinegar taste saltier? If so, you may have found a way to reduce salt without compromising on flavor.

Furthermore, the figure implies that saltiness is promoted by umami in medium to high concentrations. Adding MSG (monosodium glutamate, common in many seasonings mixes) is obviously an alternative because it is one of the ultimate umami-tasting substances. However, MSG contains sodium, which is the component in salt we want to reduce. The relative amount of sodium is lower in MSG than in regular table salt, so there would be some limited benefit so long as you don't use a comparably higher amount of MSG. An alternative that circumvents this problem is to team up with foods naturally rich in umami taste as listed in the info box.

Smell can affect salt taste. The intensity of salty taste can also be enhanced by some smells. Sometimes you may get the notion that something smells sweet. This sounds a bit odd as sweet is a taste happening on the tongue and not in the nose, and sugar has basically no smell in and of itself. As described in an earlier chapter, we make connections across the senses. This type of dialogue between the senses is the result of experiences accumulated in life. Strawberry jam is sweet and thus the smell of strawberry jam could lead to thoughts of sweetness. A very convincing research study carried

Umami-rich foods. Some foods are naturally rich in umami taste, which could also boost the experience of salty taste. Some of these contain salt, but they might still contribute to a reduction in the total intake of salt:

- Chinese cabbage, especially as fermented kimchi
- Fermented soy products, soy sauce
- Hard and stored cheeses, such as parmesan
- Green tea
- Meat, especially dried-cured ham
- Mushrooms, especially dried shiitake
- Potatoes
- Seafood, like fish sauce, anchovies, sardines, salted squid, clams, mussels, and oysters
- Seaweed, especially Japanese kombu
- Tomatoes, especially ripe or dried tomatoes ×

out by the French researcher Thierry Thomas-Danguin and coworkers showed that respondents experienced tasting water with a small amount of sardine *aroma* to taste slightly salty compared with plain water, even though neither water samples contained any salt whatsoever. And the panel experienced that water with a small amount of salt tasted saltier when a small amount of sardine aroma had been added. Some health promoters and health authorities suggest adding herbs and spices to compensate for reduced salt, but it is difficult to find solid support for this in published research. We would think that researchers in the future will find links between taste and smell, as shown in the sardine aroma experiment, to also include certain herbs and spices as salt-enhancing. But, on the other hand, why wait for the researchers when we can play with these combinations today in our own homes?

Trusting industry? Even though taste–taste and taste–smell interactions may be effective, it can be difficult to make use of them in everyday cooking because the effects may rely on tweaking very fine balances to make it work. Historically, a large part of the excess salt we ate came from industrially prepared foods and not from the meals we cooked from scratch at home. In many cases, this is still the case. In fact, according to the Norwegian health authorities (2016), a way to reduce your salt intake would be to prepare the food yourself, from scratch. However, salt from industrially prepared food is not only about TV dinners, freeze-dried soups and stews, frozen pizzas, and so forth. It is just as much about the things you put on your home-made packed lunch: pâtés, cheeses and sandwich ham. So, regardless of whether you buy your lunch at a café or you bring your own sandwich, you may have a high-salt meal. If the salt-reduction project requires that the population starts making their own cheese spreads, hams and pâtés, we are probably more in the realm of fairytales than realities. Then it would be much easier to make the industry reduce the salt content in their food, and indeed many manufacturers work seriously on this. Health authorities worldwide have for some time put pressure on the food industry to decrease salt levels. Although this approach is not straightforward, promising initiatives exist. For instance, in 1975, the health authorities and the industry in Finland agreed to reduce salt content in industrially prepared foods, and by

2002 the salt intake had decreased on average by three grams per day. Similar results have been seen, for example, in the United Kingdom. In the United States, the Food and Drug Administration have developed targets for American food manufacturers, in the European Union, a "Salt campaign" has been initiated, and the World Health Organization has its salt reduction program targeting both industry and the population in general. Still, the industry and professional food services have to make the food taste good—otherwise consumers won't buy their products. A fine balance indeed. Luckily, many of the larger manufacturers have research and development divisions that spend resources on developing low-sodium recipes. And if you can claim "less salt, the same great flavor," you may contribute to the customer's health while also being a successful food manufacturer. But while people in some countries usually bring their cold, packed sandwich lunches from home, people in other countries have their hot lunches prepared and served by professional kitchens. Thus, salt-reduction strategies also need to align with the eating habits and food culture of the population.

However, it would be rather defensive of us to sit passively and expect that the food manufacturers should solve this whole issue for us. Building on knowledge about taste, smell and connections between senses (multisensory correspondences), we may start making a difference in our own kitchens as well by:

- Experimenting with flavor combinations that could enhance salty taste: use a small amount of something sour, or try adding an umami-rich ingredient.
- Instead of adding salt, use ingredients with aromas that are often associated with salty taste, such as sardine/anchovies, low sodium soy sauce, and fish sauce.
- Use less salt when cooking and add salt at the table. However, this will give a different taste perception than a ready salted and balanced dish.
- When enjoying salt-rich food, such as a pizza, use side dishes with little or no salt such as a fresh salad. This would give a lower average salt concentration in the meal as a whole.

- Reflecting on the flavors we experience while enjoying food; it isn't snobbish to talk about what you're experiencing while eating even though some would claim otherwise.

The salt experiment. We wanted to test if some of the research-based knowledge on taste–taste interactions and taste–smell interactions could be made to work in our own kitchens. Is it possible to reduce the amount of salt and compensate with some other taste or smell? We went ahead to design an experiment that tested if adding small amounts of sour taste, umami taste, or herb aroma would make a dish taste saltier. We did two parallel experiments: one with beef bouillon and the other with fish bouillon. This way, we were able to test this in two foods that were very similar but with different flavor profiles. In both cases we made a base bouillon without salt, divided it into four parallels and added different combinations of flavor ingredients. The participants were to blind taste the samples and rank them in order of least salty to most salty. The salt level was identical in the four, near the threshold of being noticeable as having salty taste. If any of the other ingredients did actually affect the experience of saltiness, these samples should taste saltier than the one with only salt. The four parallels contained: a very small amount of MSG (parallel A), a little herb that would impart a mild aroma commonly associated with salty dishes at the same time imparting little or no taste on the tongue (parallel B), only salt for comparison (parallel C), and a very small amount of vinegar (parallel D).

Bouillons for blind tasting

A	0.3% salt + MSG (umami taste); total Na^+ is identical to the other samples
B	0.4% salt + herb commonly associated with salty taste (beef bouillon: parsley, fish bouillon: lovage)
C	0.4% salt
D	0.4% salt + 0.05% vinegar (sour)

^ Four different beef bouillons and four different fish bouillons were prepared, each with different salty-taste-enhancing ingredients, to study how the added (non-salt) ingredient affected the perception of salty taste.

The effect of tastants and aroma on salty taste

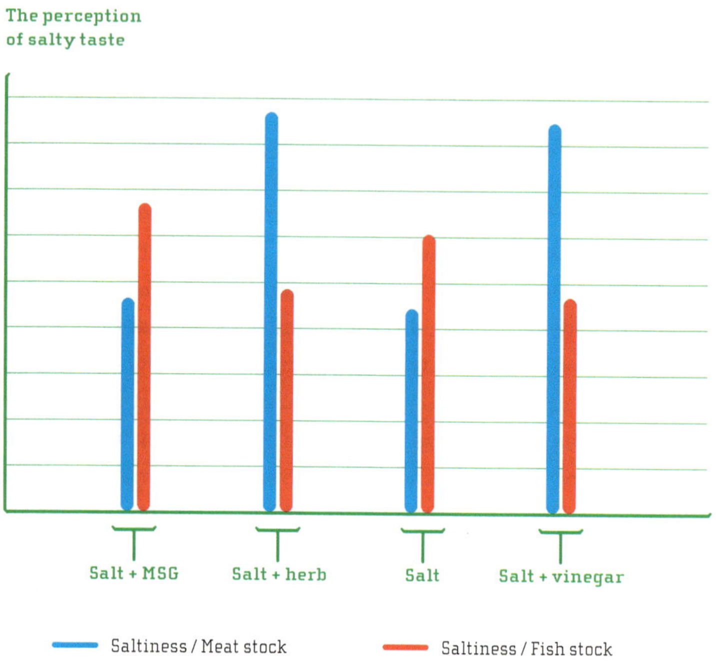

^ The samples with a mild herb aroma and sour taste were perceived as saltier in the beef stock, while the result was the opposite in fish stock.

The results are shown in the diagram. The participants tasting the beef bouillon were quite clear about the herb- and vinegar (sour) parallels being saltier than the other two: saltiness and savory aroma seemed to increase the perception of saltiness. However, the parallel experiment done with fish bouillon gave the exact opposite result. The effect may perhaps be product/context dependent. Is this effect real and, if so, how come? Could it be that the flavors of meat and fish bouillon respond differently to the added ingredients? This is not unlikely as bouillons are complex mixtures of many chemical substances that produce a complex range of flavors. The proper research experiments showing such taste–taste interactions are often carried out in very carefully controlled environments, sometimes with basic water solutions with minimal amounts of other ingredients added. In our, more relaxed and naturalistic context, subtle differences are not that easy to pick out. Maybe this is the explanation why we did experience a difference in the mild beef bouillon, but not in the fish bouillon, which often has a more prominent flavor.

Anyhow, the beef bouillon experiment tells us that there might be validity to such salt-reduction strategies, but the fish bouillon experiment shows us that they are not necessarily straightforward to apply. It certainly shows us that the world of everyday life is complex and not easily reduced to controlled lab experiments. On the other hand, it is one of the simplest of things to experiment with flavors if the only thing you need to do is to add a few drops of vinegar or some leaves of an herb to make an experiment. As long as we stay curious and continue experimenting at home and in labs, these formal and informal inquiries could maybe, in themselves, help us cut down on salt. At least, they will give us many nice experiences.

18

Physics Takes the Cake

You have invited friends and family to a celebration, maybe a birthday or Independence Day. Then you ask yourself: "Should I buy a ready-made sponge cake, or should I bake it myself?" The first option is maybe the safest bet, but the second will surely communicate more sincerely your love and appreciation for the guests. But, maybe, your experience from prior baking projects has been that your cakes don't always turn out light and airy. Last time you baked such a cake it came out flat and dense, even though you followed the recipe and treated it with the utmost care, making sure you didn't disturb or shake the light and tender sponge. You had to rush to the supermarket to get a ready-made sponge. What to do this time? The British physicist Peter Barham has a suggestion for you, claiming that his advice will ensure that your sponge cake does not collapse. Brace yourself: as soon as the cake is ready and baked in its metal spring form, you take it out of the oven and hold it 20–30 cm above the kitchen bench or a table. Then, let it fall flat down. Boom! This, he claims, will ensure that the cake will not collapse. Is he joking!? Does he expect us to trust this advice a few hours before the party? Has he tried this himself? If I try this, should I buy a ready-made sponge cake just in case? The questions and dilemmas pile up, but this claim truly begs for an experiment.

Jokes often have some core of truth, and science jokes are no exception: "If it is green or wiggles, it's biology. If it stinks, it's chemistry. If it doesn't work, it's physics." We are chemists and we're used to getting our hands dirty, but can we trust the physics professor to have actually tested his claim? How many of us haven't experienced the school science experiment that didn't work out as expected although the book said it should? With an important cake in the oven, this is not an option. So, we decided to dive head-first and test this claim in a cake-dropping experiment. Perhaps the outcome would result in some jaws dropping as well? The experiment was conducted in a food workshop with 20 participants to witness and taste the result.

The cake drop. Tatu the chef chose a very basic recipe for sponge cake: sugar, eggs, and flour. Sponge cake recipes may also contain milk, and even baking powder to help with the leavening, but we chose to keep it as simple as possible to avoid complicating factors.

The two parallel cakes were baked simultaneously in the same oven. When taken out, one cake was treated with the utmost care and placed carefully to cool in the tin. The other was held horizontally 20–30 cm above the table and dropped flat down, after which it was left to cool. When cold, the cakes were cut in half and the heights measured.

Recipe

Ingredients

250 ml sugar
250 ml eggs (6 medium sized eggs)
250 ml sifted wheat flour + some flour for the cake tin

Procedure

The sugar and eggs were beaten with an electric mixer until a light foam. The flour was added and gently folded in. The mixture was poured into a 32 cm springform cake tin and placed in a convection (hot air fan) oven for 1 minute at 200°C, and then 25 minutes at 180°C.

The recipe was repeated to give two sponge cakes.

The result. The cake that was dropped was clearly higher in the center area compared with the one that was handled with care. The latter had the typical high circumference and pit in the middle, while the one that was dropped had a more even height throughout. In the middle area, the one that was dropped was approximately 20% higher than the one handled with care—quite a spectacular result! Well, not that surprising as we had secretly pretested this in advance. We had safeguarded ourselves and did a corresponding experiment at home, one month earlier. In this experiment, the result was much the same, but with a somewhat less marked difference. In that case, we used some baking powder in the recipe, so the cake handled with care came out quite tall as well, but the one that was dropped came out 15% higher and even had a center area that was higher than around the rim.

We are quite convinced that the British physicist's claim is indeed correct, but what can be the explanation for this apparently counterintuitive result? This requires a closer look at cake structure and going back to our structural friends, the dispersions (as described in the sausage chapter).

What makes a cake rise? Structurally, a sponge cake is a solid foam—a solid substance with gas bubbles distributed throughout the structure. Just like the inside of the pillow you might be sitting on right now while reading this book. If we simply stir together egg, sugar, and flour and bake it in the oven, we would get a pancake; flat and dense. The foamy cake structure is achieved if we beat eggs together with sugar to produce a liquid foam, full of air bubbles, and then gently fold in the flour to avoid too many bubbles bursting. We still have a fragile liquid foam, but somewhat more thick and viscous. When the cake enters the oven, it rises—even though no baking powder or other leavening agent has been added. The reason for this is that the water from the eggs in the batter evaporates. Liquid water turns into steam. Steam—water in the form of gas—requires more than 1000 times the space compared with liquid water. One teaspoon of water turned into steam would, in the ideal case, expand to more than 5 liters. So, when the water evaporates in the liquid batter, it ends up in the bubbles, who in turn expand radically. The whole cake increases in volume and therefore rises, the internal vapor pressure in the cake keeping it up, competing with gravity, which constantly

^ **Liquid foam:** Chemically speaking, a cake batter is a liquid foam. Water (blue background) is the continuous phase, and air bubbles (white) are the dispersed phase. Proteins (in particular from eggs, orange and purple) and sugar (blue) are dissolved in the water. Starch granules from the flour (yellow) float around as solid, dispersed particles.

pulls the whole cake down. The air that we beat into the batter would also expand somewhat when heated in the oven, but this expansion is not much more than 25%, and expansion of air, therefore, has a much smaller effect than expansion of liquid water into steam.

In the early stages in the oven, when the cake is still practically a liquid foam—bubbles dispersed in a liquid—it is very fragile and sensitive to mechanic disturbance. If this wobbly mass gets a nudge, the bubbles of hot steam maintaining the expanded foam might find their way up and out, even leaving behind channels for more steam and gas to escape. This is the moment where the cake requires peace and quiet. However, after a while, the batter solidifies due to chemical reactions facilitated by the high temperature: gelling of starch from the flour and coagulation of proteins from eggs and flour. The liquid foam turns into a solid, but soft and brittle, foam. The cake is still in the oven, but, in addition to the internal pressure of the steam, the solid structures of gelatinized starch and coagulated proteins now hold it up. So now you may safely dance and jump around the cake as much as you wish without the fear of it collapsing.

Then the moment arrives when the cake is ready to be taken out from the oven. How come your beautiful fluffy creation shrinks in front of your very eyes, the high peak in the middle turns into a valley, and the smooth surface of the cake is all wrinkled? The sponge still contains a considerable amount of water. When the cake is taken out, the temperature soon sinks to below 100°C, and the steam still inside the recently expanded bubbles condenses back to liquid water. The recently expanded bubbles experience a vacuum effect; they are sucked together from the inside. What looks like a cake "falling together" is actually a mild implosion. However, the cake is now a solid foam that is more rigid than it was as a liquid foam prior to baking. If we were to bake a bread, the solid foam would be rather strong and elastic, and the crust so sturdy that the bread would not contract that easily, while in a sponge cake the structure is still rather soft and brittle and prone to collapse. Channels that we so much feared would let steam escape if the cake was disturbed in the early stages of baking would now have been much welcome to let air in from the outside as steam condenses and bubbles contract.

The oven baking process

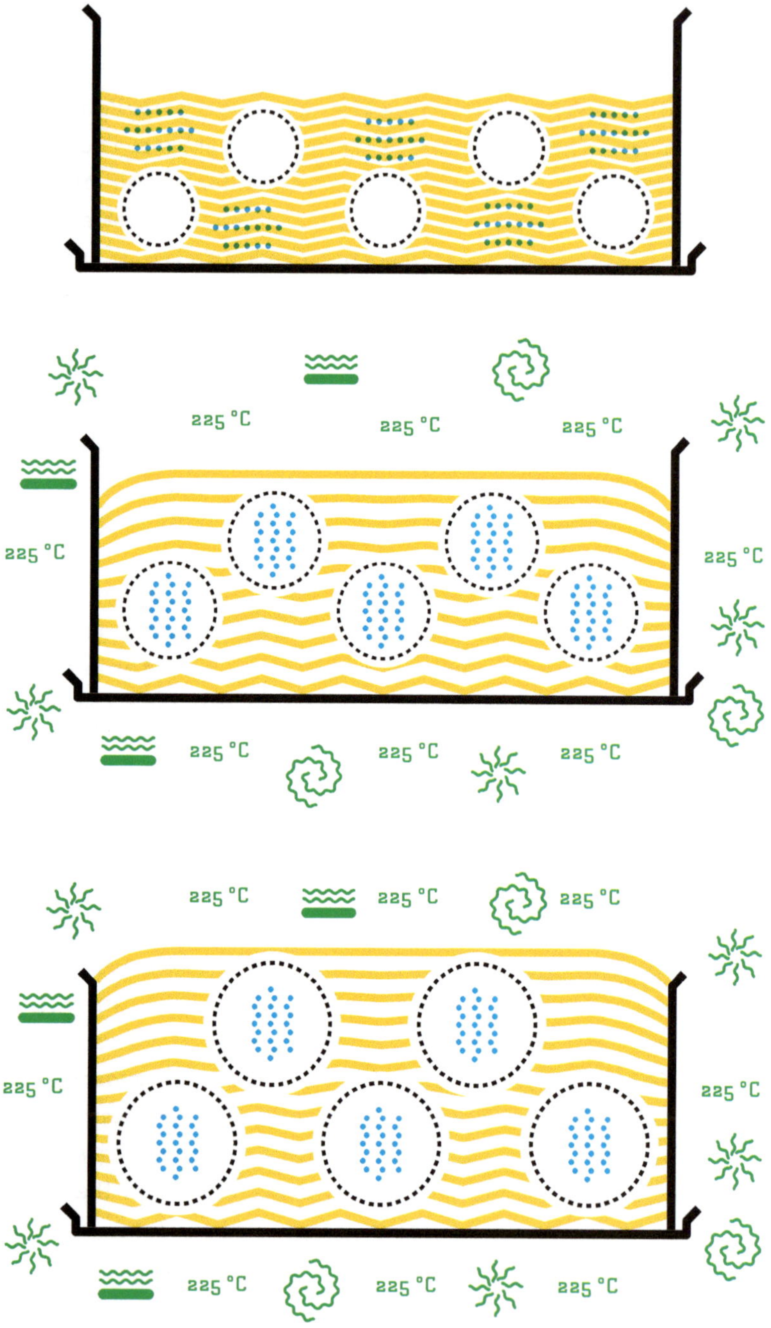

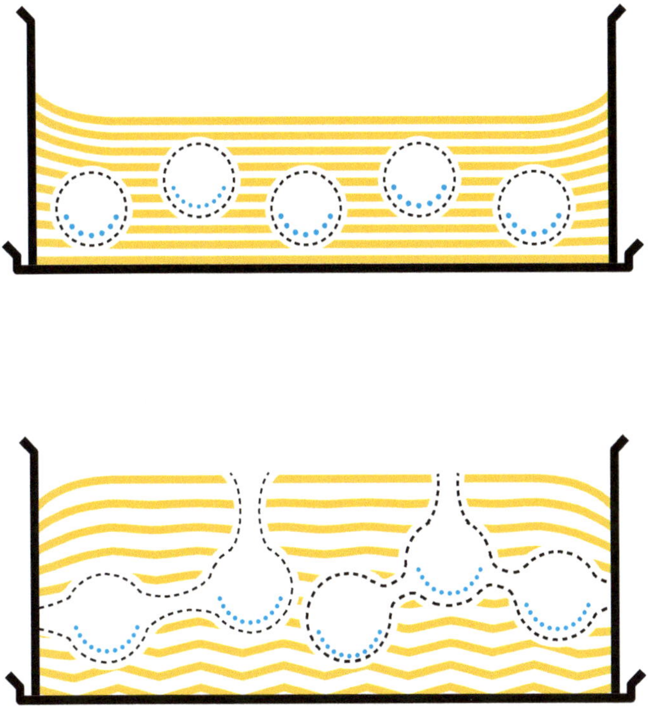

^ When water in the cake batter boils as a result of heat transfer from the oven, the air bubbles are filled with steam and expand (left-hand page, from top and downwards). When the cake is taken from the oven, the steam condenses into liquid water. The bubbles—and the cake—contract (right-hand page, top). However, if the cake is subjected to a shock just after baking, cracks will form between bubbles and on the surface of the cake. Air can then pass through these new channels to replace steam that has condensed into water (right-hand page, bottom).

What happens when the cake is dropped? We do not know whether Barham's advice came after an accident in the kitchen making him reflect upon this, or if the hypothesis came before the experiment. Our tests anyhow support his claim and indicate that he might have found a good way to ensure a high sponge. His explanation for why this works goes as follows: when the cake is taken out from the oven, it is a solid foam: lots of bubbles surrounded by a fragile solid dough (crumb). If this structure receives a physical shock, cracks will appear between bubbles, and on the surfaces of the cake functioning as ventilation channels. Small and rather firm bubbles become interconnected, united into a more open network. Additionally, the steam inside the cake condenses and contracts, but instead of the bubbles contracting with the steam, air from the outside can be sucked in through the cracks to fill the bubbles. Even though the crumb and crust are soft and fragile, they are strong enough to maintain the structure. And you will not have to make double the amount of cream to fill a pit in the middle of the cake.

As to yet, we have not found any scientific studies describing such shock treatments of sponge cakes, so Peter Barham's explanation should perhaps still be considered as a hypothesis. But it would be interesting if research could document the effect of the cake drop. Pastry chefs and bakers might have a safe route to high cakes, they could make their cakes 10–20% higher without increasing the amount of ingredients, and the customers would have cakes with fewer calories of energy.

This time we must admit that even though it is physics, it still works in real life and not only under idealized and controlled conditions. At least, the times we have tested it. The question remains: would you put your trust in the physics if there was an important occasion at stake...?

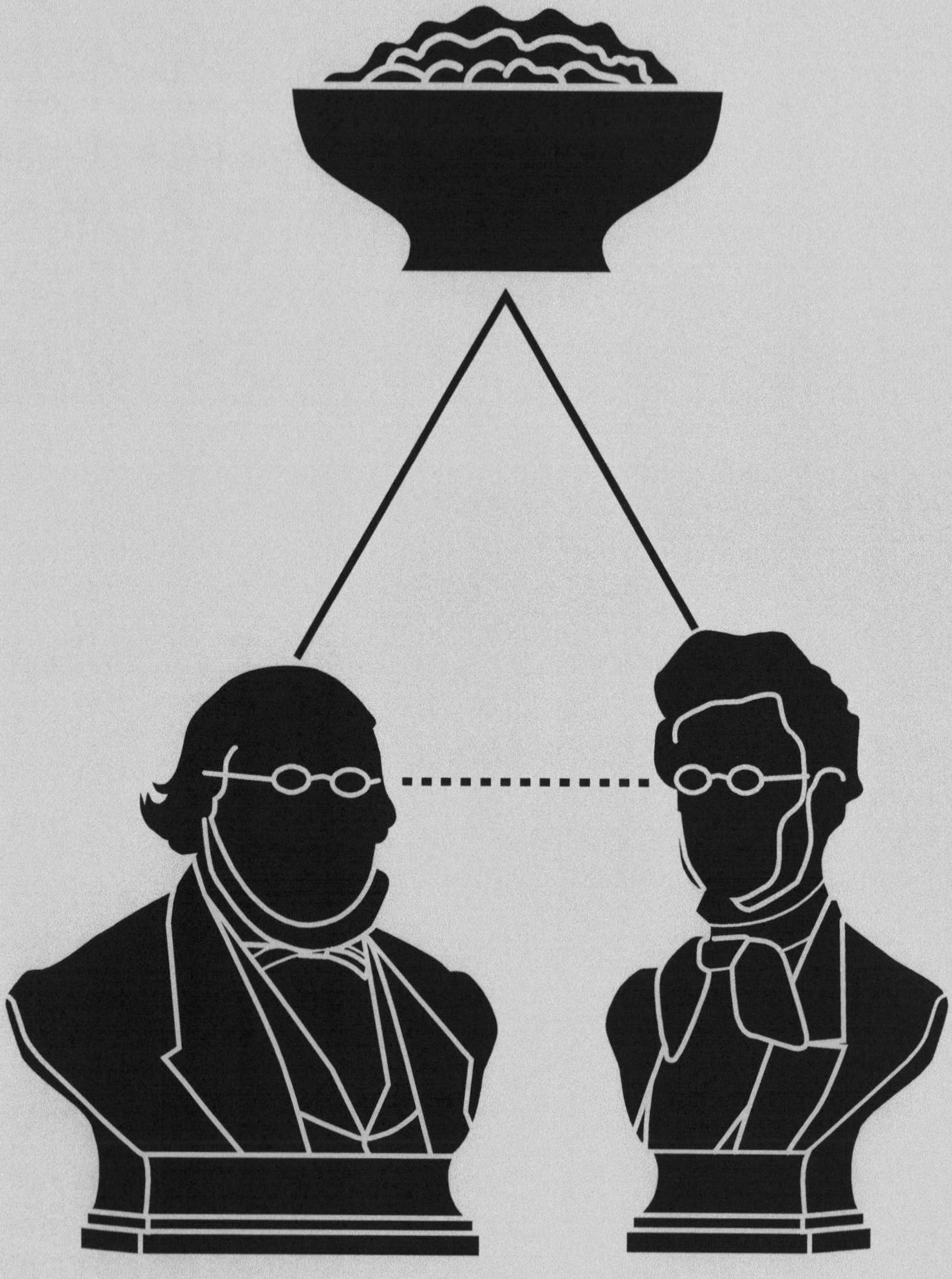

19

The Great Norwegian Porridge Feud

In 1864, a rather unusual cookbook was published in Norway bearing the name *Sensible Cookery—A timely cookery—and household book*. The author was a mysterious person hiding behind what was obviously a pseudonym: Clemens Bonifacius, translated from Latin meaning "The gentle helper." The book consists of two main parts. The first is a general text on nutrition and home economics, while the second part is a collection of recipes and cooking procedures like most other cookbooks. If you were a farm wife or housewife reading this book, you would probably not experience the first part as very gentle, but rather harsh and scolding. Perhaps you would think that a more proper pseudonym for the author should be, "the strict educator," your thoughts wandering to the time when you went to school and the teacher used the rattan stick on your hand if you hadn't done with your homework. When opening the book, you would read that the way you cooked your porridge was entirely wrong; it was against the knowledge of modern science. And cooking porridge was not a small issue in those times, as two-thirds of the people in the relatively poor Norway (100 years before the faucets of North Sea oil started flooding) ate porridge three times a day, 300 days of the year. By today's standards the porridge of 19th century Norway was not a very luxurious dish: coarse whole meal flour, water and a pinch of salt in a pot cooking over an open hearth.

"Mr. Bonifacius" had been travelling in central Europe, especially in Germany, to pick up the most recent scientific knowledge. He claimed that as much as one-sixth of the flour used by Norwegian farm wives for porridge was wasted. If true, this was a big deal, as he calculated that it would amount to a loss that would be clearly visible on the national budget. If only the farm wives would do things correctly, they would help their families to a more viable economy in an otherwise poor country of mostly peasants.

What was the problem, and who was C. Bonifacius?

Bonifacius criticized the farm wives' cooking practice of being silly, old-fashioned, unscientific, and inefficient. The two prominent examples he used were how they cooked their porridge and how they roasted their coffee (it was common to buy the coffee as unroasted beans and roast one's own coffee at home). A typical procedure when cooking porridge would be as follows, based on 5 dl whole meal barley flour:

Heat 2 l water to the boil in a pot and add flour little by little while stirring until you have added around 4 dl flour. Let the porridge simmer for 20 minutes. Take the pan off the heat and add the rest of the flour in three rounds, stirring vigorously between each addition.

Leaving some of the flour until after the cooking was called to "pound" or "pestle" the porridge or pounding/pestling flour into the porridge. According to Bonifacius, this way of cooking was contrary to scientific knowledge because the flour added in the end was not cooked properly and would therefore not be digestible. Starch must be heated in water to swell and become available for people's digestive systems. If as much as one-sixth of all flour for porridge was to go through the families' bowels undigested, that would be a great loss to the country! Bonifacius literally wrote that "the women's order in this country is truly such that they have to learn everything anew" and "our farm wives can't even cook a porridge properly."

The book was well written and sold out within the first year. It also aroused attention from scholars, particularly from the now famous

"The women's order in this country is truly such that they have to learn everything anew"

"Our farm wives can't even cook a porridge properly"

"After all, the physicians are the experts here."

— Peter Christen Asbjørnsen

"They have cooked porridge day in and day out for a thousand years or more"

"It could not be the conclusion that they were not able."

— Eilert Sundt

sociologist Eilert Sundt, who had travelled across the rural areas of Norway to learn and understand the ways of living among the common Norwegian. When it became clear that the person hiding behind the pseudonym was the famous natural science scholar and folk tale collector Peter Christen Asbjørnsen, the stage was set for a feud in the public media between the natural scientist and the social scientist. Based on his own observations and documentation, Sundt rebutted Asbjørnsen's harsh claims about Norwegian cooking practices with

Typical 19th century porridge from Norway

Ingredients

270 g whole meal barley flour
2 l water
1 tsp salt

Accompanied by

Sour milk (buttermilk)
A small knob of butter, or syrup and some Norwegian brown whey "cheese"

Procedure

Heat the water to the boil in a pot and add flour little by little while stirring until you have added about 5/6 of the flour. Let the porridge simmer for 20 minutes. Take the pot off the heat and add the rest of the flour in three rounds, stirring vigorously between each addition.

To serve, you need two bowls. One larger bowl for the porridge with a small "butter eye," or a small eye of brown whey cheese and syrup. The other smaller bowl contains the sour milk. The porridge is eaten as follows: you take a spoon of porridge, dip it in the "eye", and then dip it into the sour milk. If the whole party shares the two bowls using only their individual spoons, you are close to the authentic way of eating this porridge. After the meal, you lick off your spoon thoroughly and hang it back on its regular place on the wall.

>>>

Norwegian "brown cheese"
This is a Norwegian delicacy, and an everyday food. Technically it is not cheese, but it is boiled down whey with some added cream, which attains a brown color with a caramel-like flavor and the structure of a somewhat brittle cheese. Hence, it is perhaps even more correctly described as a type of unsweetened caramel than a cheese. Since whey, the liquid by-product from cheesemaking, contains both proteins and sugars and the product is made by extended boiling down until it becomes solid, it is among the ultimate examples of a product of Maillard reactions. Also, this way, all the components of milk are utilized to make food, both the curd (cheese proteins) and the whey, that which is left when the curd is taken out. ×

the counterclaim that these farm wives had kept their households running, and cooked their porridge, for a thousand years. Surely, their experience and everyday observations would have guided them to good and efficient ways of doing things! So, this was not only a debate between different sciences, but also a battle between science and craftsmanship, between tradition and economy—and between national identity and foreign influence.

Who was right? A couple of years after the publication of *Sensible Cookery*, a third person entered onto the stage. Frans Christian Faye, medical professor at The Royal Frederik's University (now University of Oslo), held a lecture presenting empirical evidence apparently refuting Asbjørnsen's hypothesis. He had carried out an experiment, today often done as a classical science experiment in schools: iodine is added to various foods in order to detect starch by staining the starch granules thus making them visible in a microscope. Faye had placed his research assistant on a porridge-based diet, cooked with "late addition of flour" to mimic the farm wives' practice. The assistant was served porridge three times a day for six days, adding up to nearly 3.5 kg flour in total, followed by analysis of stool samples from the assistant. However, they were not able to find any detectable starch, so they concluded that cooking the porridge according to the procedure used by the farm wives did not make any difference; the starch was digested anyhow it was cooked. Thus, Asbjørnsen's claim was seemingly disproved and Sundt could breathe more easily. On the other hand, Asbjørnsen's import of foreign and modern science still had an important impact because this is said to have spurred the initiation of nutrition research in Norway. So, ethnologist Andreas Ropeid, who has written in the highest detail about this historical issue, concludes that Asbjørnsen and Sundt both got their way, and the feud had a happy ending.

Why did the farm wives cook their porridge this way? Looking in recipes of today there are few, if any, traces of the 19th century practice of leaving part of the flour to add in the end. So why did they do this? In historical accounts of the porridge feud, the farm wives' own voices are seldom heard, except via the statements by sociologist Eilert Sundt. Did anyone ask the cook directly why she cooked the porridge this or that way? Sitting at a cafe table in Helsinki with our respective bowls of porridge, we wondered why on earth the Norwegian farm wives cooked their porridge the way they did. We have tried to reason and have three possible hypotheses: the nutrition hypothesis, the texture hypothesis, and the flavor hypothesis.

The nutrition hypothesis. Doctor Faye presented one possible interpretation of the reasons for this cooking practice: if the porridge was cooked in a way that it was not as easily digestible, but took longer to digest, it could be beneficial for the working man. The food would stay longer in the belly, the food would provide energy over a longer period, and worker or farmer would not become hungry too soon. So, this way of cooking the food would be suitable for body laborers, which constituted the major part of the Norwegian population at the time. Today, surrounded by abundance and constantly being worried about getting too much carbohydrates from our food, we might actually be happy to go back to pre-Asbjørnsen times. While he was worried about people wasting food as part of the energy-containing nutrients staying in an unabsorbed form (not bioavailable), many today would be happy if the abundance of food we ate liberated less energy into our bodies. Indeed, one of the wonder substances of today for those who want to watch their weight is resistant starch. These are types of starch that are non-digestible, or slowly digestible, in the small intestine where nutrient absorption takes place. Rather, they are broken down by bacteria in the large intestine and thereby function as fiber, without providing any energy. Today, we also know that starch is bioavailable only in gelatinized form, as it is generally after cooking. However, if the gelatinization does not take place, or if the starch is recrystallized, it will not be digested. This recrystallization takes place, for example, during refrigeration of cooked potatoes or pasta. The Norwegian farm wives might have been ahead of their time and served porridge where part of the starch from the flour was in its resistant form, functioning as fiber, and thus keeping the hunger at bay for a longer time. Furthermore, this recent knowledge proves Asbjørnsen right.

The texture hypothesis. Historian Jarle Sanden at Romsdal Cultural Historic Museum in Western Norway hints toward a rather different reason for adding flour after cooking, as he uses the term "really hardly pounded/pestled porridge" in a book about food traditions in the Romsdal region of Norway. Leftovers of such firm porridge could, the day after, serve as a replacement for potato when sliced and fried in bacon fat. Indeed, it was common to cook enough porridge for more than one meal, and a usual evening meal was lumps of porridge reheated in milk or buttermilk. Food writer Catherine Brown describes a similar practice from Scotland, another country where porridge has been an important staple. Porridge leftovers, if they are firm enough, serve well as a packed lunch for the working farmer to bring with him out in the fields. We have ourselves corresponded with two informants of age, asking them if they have ever experienced this way of cooking porridge and if they could explain why it was done. One of the informants used the same texture explanation as Sanden: "Here is a true story about the porridge: in old times, our neighbor cut grass all the way up to the foot of the mountain, and it was important to be meticulous about it so there would be enough feed for the livestock. Of course, they brought porridge in a porridge pail [small wooden bucket with lid]. The pail tipped over and the porridge went rolling down the hill. That porridge was pounded for sure!"

The flavor hypothesis. Historical literature describes porridge as an important source of energy for people whose lives were characterized by hard physical labor. Nowhere in the literature do we find any statements about taste or enjoyment of this food. Is it so that food was seen only as fuel for working bodies except during festivities? Were flavor preferences pushed in the background by necessity in a country inhabited by hard-working peasants? Surely, the cook would try to make the food as enjoyable as possible in this context, and not only in the wealthier contexts of traders and aristocrats? This porridge, literally translated in Norwegian as "water porridge," does not enjoy a reputation for being among the culinary high points in Norwegian cuisine; although, some would say that this reputation is somewhat unfair. Making the food more palatable is probably always a motivation for anyone making staples, and if we look to Finnish food culture, this is exactly where we find a hint at making the porridge tastier. Looking to Finland is not entirely unlikely since there was a great migration from Finland westwards to Norway during the 16th and 17th century. Maybe a procedure or recipe from Finland might have found its way to Norway? The practice of sweetening starchy foods by taking advantage of naturally occurring enzymes in raw materials is an old practice in Finnish food culture (but virtually unseen in contemporary Norway). Such sweetening can, for example, be achieved by adding non-heated flour to starch-rich food after it is cooked. This still-existing common practice is found in a traditional Finnish dish, the Christmas dish called "imelletty perunalaatikko"—potato casserole. It is made with potatoes, water, and flour in a manner that in principle might resemble that of the Norwegian porridge. Potatoes are cooked, mashed, and left to cool slightly. When lukewarm, some flour is mixed in and the casserole left at room temperature or slightly above for several hours, preferably overnight. During this time, amylase enzymes from the flour will break down the starch abundant in the potatoes into sweet sugars. The result is a looser texture, because some of the long starch molecules that otherwise will be entangled in each other, are broken down into small water-soluble sugar molecules. Starch does not taste much, but sugars are sweet. By using the naturally occurring enzymes in flour, the dish becomes sweeter without adding any sweetening. When the dish is then baked in the oven, the presence of sugars results in browning reactions to give a light brown

Finnish Christmas potato casserole (imelletty perunalaatikko)

The parts of the recipe that concern enzymatic sweetening, and what to do if this fails, have been italicized.

Ingredients

- 1 kg potatoes (starchy/mealy variety)
- 1 dl wheat flour
- 25 g butter or margarine
- 0.25–0.5 l milk
- 1 tablespoon syrup, if necessary
- 1.5 teaspoons salt

Procedure

Peel and cut the potatoes into cubes. Cook in a small amount of water until soft and strain. Mash the potatoes and *let cool until approximately hand warm*. Mix in half of the butter and *a little more than half of the flour into the potatoes*. Add the rest of the flour on top. *Keep the pan covered overnight, or at least 5–6 hours, in a warm place*. You may stir it once or twice, but this is not strictly necessary. *Taste every once in a while to monitor the sweetening process. If the texture has not become looser during the first 6 hours, don't worry, just add a spoonful of syrup.* Add cold milk to the mixture and stir thoroughly, *the mixture should be loose* like porridge, much looser than regular mashed potatoes. Add salt. Pour the mixture into a high-walled ovenproof dish and add the remaining butter on top. Bake for 30 minutes at 200°C to bring the mixture to the boil. Reduce the oven temperature to 150 °C and continue cooking for another 2 hours until you get a *slightly sweet and light brown dish with a dark brown crust*.

dish with a dark brown crust, very different from what you'd get if you put the mashed potatoes straight into the oven. This sweetening process is sometimes quite tricky to achieve, so recipes for the potato casserole often include a safeguard: if it hasn't become sweet by its own, add some syrup before placing it in the oven.

Such starch-cleaving enzymes are found several places, including in our own saliva. This principle is used to produce traditional alcoholic drinks such as the Japanese sake and chicha found in Central and South America. In the latter case, the starchy food is chewed and then put back into the tray. The enzymes from the saliva break down the starch into sugars whereby naturally occurring or added yeast may digest these sugars to give alcohol. You can test this phenomenon yourself by taking a teaspoon of flour in your mouth and leave it there for a few minutes; you should notice that it turns slightly sweet. The knowledge about enzymatic cleavage of starch to sugars by amylase enzymes from saliva was already known during the time of the porridge feud, and indeed described by Dr. Faye. However, none of the literary sources say anything about a process of using uncooked flour as a source for enzymes to sweeten the porridge. Could it still be that the Norwegian housewives had learnt this from the Finnish and applied it into their own cooking practices? After all, a sweeter porridge might be more enjoyable, and maybe they could cut down on the expensive sweet syrup or sugar? They didn't need to know about the chemical processes going on between enzymes and starch; the only necessary knowledge would be to know how to do it and the fact that the porridge would become sweeter.

We went ahead to test whether it would be possible to cook a sweet porridge using the old method. In the context of a science teacher conference in Norway, we cooked two parallel portions of porridge. One was cooked according to modern recipes, as recommended by Asbjørnsen: water was brought to the boil and all the flour was added gradually while stirring. The porridge was thereafter boiled for 20 minutes. The other parallel started four hours earlier. With the exact same amounts of flour and water, the porridge was cooked according to the farm wives' procedure so that the last one-fifth of the flour was added after the porridge had cooled to below 60°C. Above this temperature, amylase enzymes are deactivated (denatured), so we kept the porridge warm, but below 60°C, for four hours. The two porridges were served to ca. 45 science teachers in a blind tasting, and the participants were asked to rank them according to sweetness. The result was very convincing indeed: the "modern" procedure received only one vote for being sweetest. The remaining 44 or so participants experienced the porridge cooked according to the farm wives' practice as sweetest. We concluded that our informal experiment had quite overwhelmingly strengthened our claim that it is possible to cook a porridge with increased sweet taste using a procedure that resembles that of the 19th century farm wives.

Can we prove, or falsify, any of the hypotheses? We turn to the science philosopher Karl Popper, who wrote that it is not possible to prove a hypothesis to be true. The best we can achieve is to falsify possible hypotheses, and those that still stand after we, by the very best of our ability, have tried to falsify them, are the ones currently valid. In this case, it is difficult to falsify any of the three hypotheses. Cooking porridge the 19th century way is a human practice that is no longer present in today's food culture, at least not to our knowledge. All three hypotheses are possible explanations, and it could even be that all three are correct.

But we would call to attention the fact that the texture hypothesis and the flavor hypothesis, to a certain extent, contradict each other, because sweetening the porridge by enzymes would also give a much looser texture. You wouldn't get a porridge the next day that was both sweet and could roll down the hill in one piece; sweet and firm at the same time is an unlikely combination. But an important factor is that this sweetness is quite difficult to achieve, the exact reason for the Finnish potato casserole recipes including syrup as safety net. We might speculate that some housewives were able to make the sweet porridge while others failed. But in failing, they got a porridge with a firm texture which gave another benefit. You lose some and win some. This is, of course, speculation, and since rural cooking practices are seldom documented in writing, it is not possible to trace the motivations of the farm wives' cooking practices. Although many were literate in 19th century rural Norway, the fact that these people apparently didn't find their everyday activities important enough to document was the very reason that Eilert Sundt made his travels around to talk with them and observe how they lived.

So, what have we gained by reopening this chapter in history? Have we just added more confusion by introducing yet another hypothesis? Yes, indeed. But that is also how knowledge is built. And we noticed that those previously studying the porridge feud had apparently not considered, or at least described, flavor to be a motivation for how people cooked their porridge. Surely, trying to make something you eat 300 days a year a bit tastier could be a possible reason for doing things a specific way?

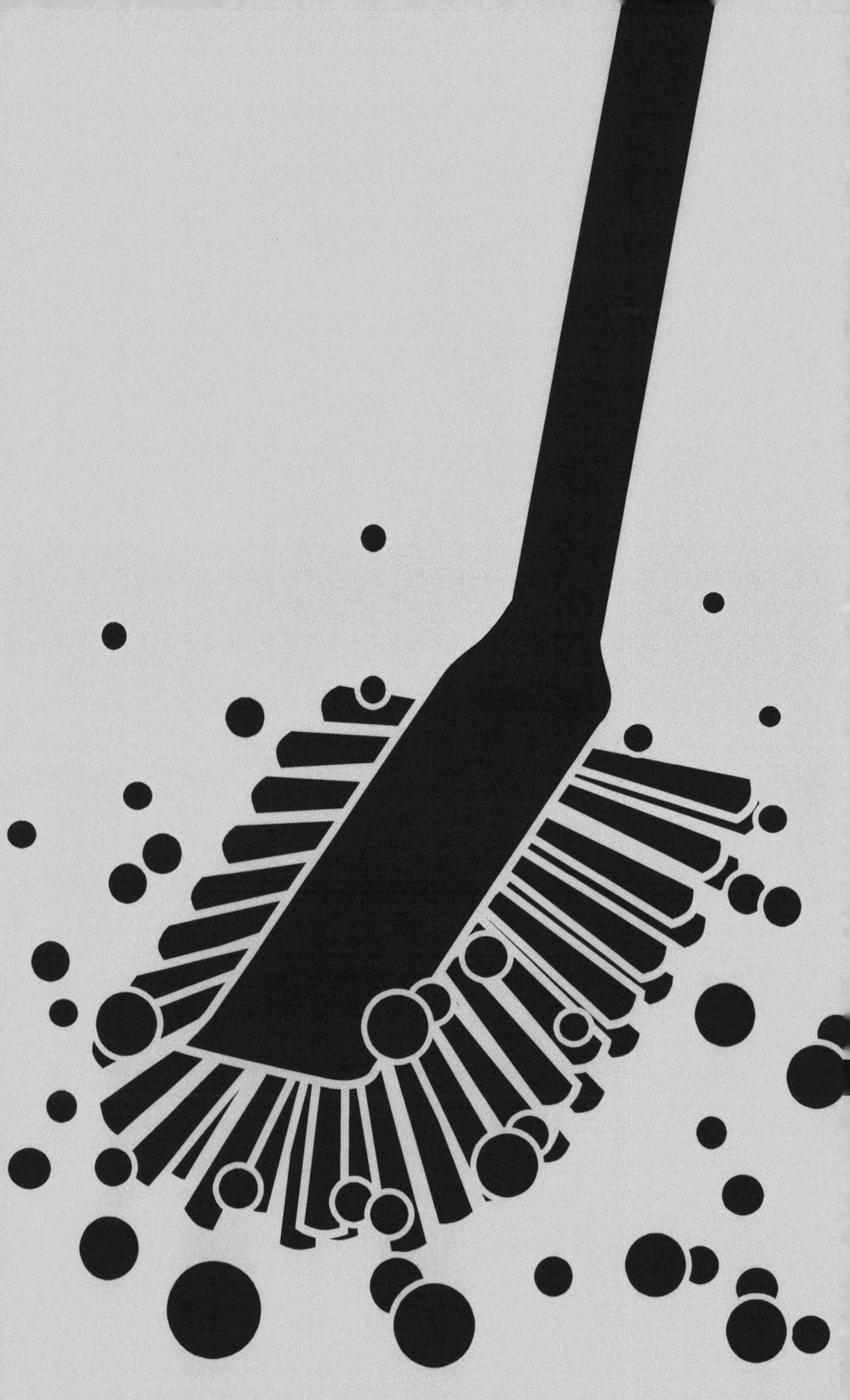

20

Doing the Dishes

The meal is finished, and it is time to do the dishes (any volunteers?). After all this cooking with science, we would hope that chemistry could also help us clean up the mess. Science is a field of classification, so let's start the dullest part of dining by classifying the dirt from dining. Food scraps can be seen as consisting of four main types of substances: fat and grease, proteins, carbohydrates, and burned residues in your pots and pans. And you have four different strategies to get rid of these sticky fellows, namely rinsing, dissolving, scrubbing, and breaking them down chemically.

Fat and grease can be removed with traditional soap, those split-personality molecules with both a hydrophilic (water-loving) and lipophilic (fat-loving) part. The soap molecules are not truly dissolved in the water but rather form nanoscale particles, micelles, which are easy to rinse away. The lipophilic part aligns toward the fat, surrounding the microscopic (or nanoscopic) fat droplets while the hydrophilic part faces out toward the surrounding water. Thus, the soap molecules facilitate the marriage between the fat from the food and the dishwashing water that runs past, dragging some fat along with it down the drain.

In dishwashers, a very different strategy must be applied for at least two reasons: firstly, the machine would face serious problems pumping and spraying soapy foam, and secondly it isn't able to physically scrub the dishes. Instead, the detergent contains a basic (alkaline) substance that reacts chemically with the fat, basically converting it into soap, which is then rinsed away with the hot water.

^ Soap molecules (black) have a lipophilic (fat-loving) and a hydrophilic (water-loving) part. The lipophilic part aligns in toward the fat, the hydrophilic toward the surrounding water. This way, the fat droplets follow along with the water.

So, dishwashing detergents don't contain soap, but some soap is produced along the way in a chemical reaction type called saponification. So, while hand-washing involves quite a lot of physical force combined with forming of dispersions, mainly foams and emulsions, machine-washing relies to a greater degree on chemical reactions between detergents and dirt.

Proteins are large molecules, at least in the world of molecules, that are easily stuck to surfaces. When uncooked, they are dissolved in the water of the food, such as in a raw egg, and will therefore easily find their way into cracks and scratches. When the proteins are heated during cooking they coagulate and solidify—basically functioning as a glue that fixes the food to the surface. Scratches on the surface of kitchenware might be smaller than visible to the naked eye, even microscopic, but still large enough for proteins to have ample space to get a firm foothold.

When you use fat for pan frying, the fat fills these scratches and, to a certain extent, obstructs proteins and other molecules from finding their way in. Coagulated proteins are not water soluble, and soaps or bases will not dissolve or break them down. Earlier, scrubbing was the most effective way to get rid of proteins, but the downside of scrubbing is that you scratch the surface and the kitchenware will be even more prone to sticking the next time you cook.

Today, dishwasher detergents contain protease enzymes, a class of enzymes that break down proteins chemically, turning them into water-soluble and rinseable substances, or simply break them down enough for the dirt to come loose from the kitchenware. Indeed, this is the same class of enzymes also found in fruits such as kiwi and pineapple, making your mouth sore when you eat too much of it or eat it slowly because they start reacting with the proteins in your skin to break them down. They are also found in carnivorous plants, allowing the plant to dissolve and make digestible the unfortunate insect that might fall in the "enzyme bath" usually found in the flower. And proteases function to tenderize meat as they break down muscle proteins, be it as naturally present enzymes doing their work during curing of meat or as deliberately added tenderizers in fresh meats.

For a frying pan with stuck remains of an omelet, soaking it for extended time in warm water with some dishwashing powder might do the trick, if you have the time to let the proteases do their job. If you are a busy restaurant chef, scrubbing might be the only alternative. If so, using a scrub that is softer than the surface of the kitchenware is a good bet, as it would reduce the risk of further scratching.

Carbohydrates are often not that difficult to get rid of. Many of them, such as sugars, are naturally water soluble. Larger carbohydrates such as pectin and cellulose do not burn and stick to surfaces as easily as proteins. The most challenging carbohydrates are perhaps sticky, gelatinized starch from starchy foods such as potatoes, cornstarch, rice, and flour. The long polysaccharide molecules dissolve in hot water and turn into hardened glue-like substances when drying. As for proteins in machine dishwashing, this challenge is solved using enzymes. Indeed, this is the same class of enzymes as those that turn starch into sweet sugars described in the porridge chapter describing

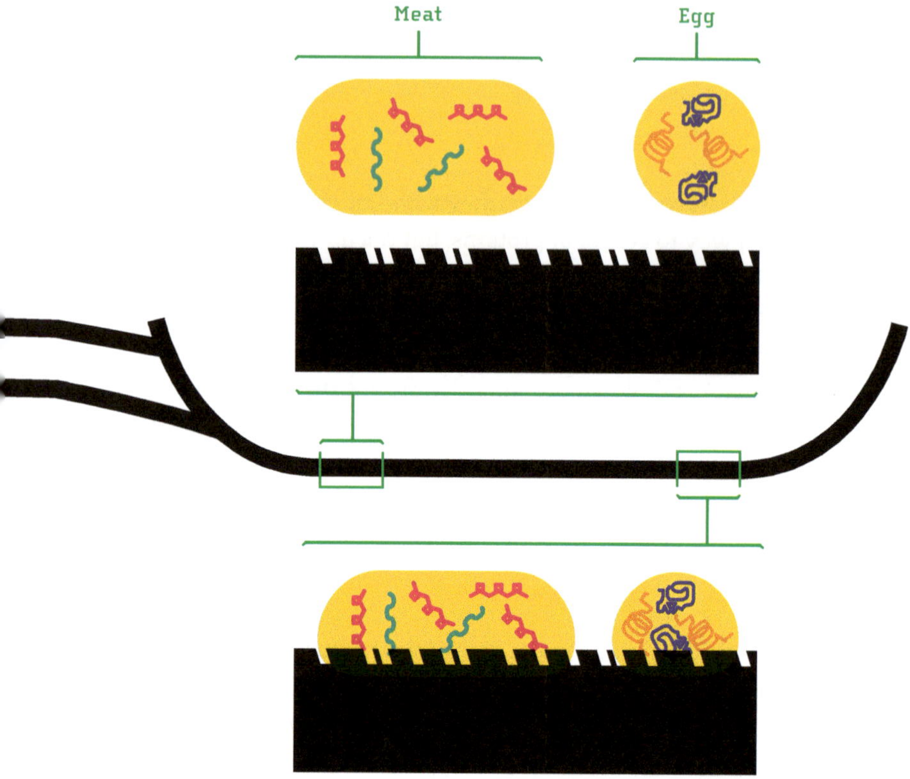

^ **Food scraps in small scratches:** Proteins and carbohydrates dissolved in the water from the food find their way into microscopic scratches on the surface of the kitchenware. During heating, the proteins coagulate and the carbohydrates gelatinize, thus making the food stick to the surface.

enzymatic sweetening and the Finnish Christmas casserole: amylase enzymes. For the home cook who faces dishes with stubborn starchy residue, the strategy to clean is the same as for proteins: chemically by soaking in water with some dishwasher detergent, or, if you are in a hurry, using your muscles.

Burnt residues are a mixture of all of the above that have been subject to such high temperatures that they are partially turned into carbon to give dark stains. As we have all experienced, these are hard to get rid of and often require some physical force or harsher chemical treatment, such as soaking with dishwasher powder. To deal with dark stains, older detergents contained chlorine bleach, whereas modern versions may contain substances that accomplish two things at one time. When dissolved in water, the salt sodium percarbonate produces both hydrogen peroxide, "active oxygen," and carbonate ions at the same time. The former bleaches dark stains by chemical reaction, while carbonate ions make the conditions basic (alkaline) to turn fats into water-soluble soaps. Additionally, these modern detergents do not pollute because the peroxide is broken down to oxygen and water, and carbonate is among the most abundant ions in nature. So, the burnt dirt indeed challenges the designers of dishwashing detergents because it requires the coordinated efforts of all the tools in the detergent designer's toolbox. And you might even need to do some scrubbing on top of that. Detergents for manual dishwashing are usually not that complex, and this is probably the reason that these don't work as well for soaking your ovenproof dish after a lasagna dinner.

Anyway, no matter how you wash your dishes, it is nice that they are now clean and waiting for the next cooking session. Theoretical cooking, such as reading this book, has the benefit that it doesn't leave dirty dishes; a great advantage over real-life cooking. The obvious disadvantage with such theoretical cooking is that you often might end up rather hungry. Our recommendation: a healthy portion of both!

Glossary

Acid: In a chemical sense, the opposite of a base. When an acid is mixed with water, the solution (which is consequently acidic) has a pH below 7. Acids have a sour taste.

Amino Acids: Amino acids are organic chemical compounds that can form chains, or polymerize, into larger molecules. Proteins consist of long chains of amino acids.

Amylase: Enzymes that can cleave starches into simple sugars such as maltose and glucose. In plants, amylases break down starch into sugar that plants in turn use for growth and energy production. Amylases occur in saliva and are an important part of the digestive system in humans.

Antioxidant: Chemical compounds that prevent oxidation of other compounds. In food, antioxidants are used to prevent the oxidation of fat. In biological systems, antioxidants inhibit processes caused by oxidative stress, among others associated with a number of diseases.

Aroma: Related to sensation of odorant molecules, olfaction, sensed by the scent receptors in the nose. A requirement for a compound to have aroma is that it is volatile. It must be able to evaporate to reach the nose either through the nose (nasal cavity, called orthonasal olfaction) or through the oral cavity (mouth, called retronasal olfaction).

Ascorbic Acid or Vitamin C: One of the water-soluble vitamins. In food manufacturing, ascorbic acid and its salts are used as additives to inhibit oxidation reactions that lead to food spoilage (additive codes 300 ascorbic acid, 301 sodium ascorbate and 302 calcium ascorbate). Ascorbic acid is found naturally in many fruits and vegetables.

Astringency: A sensation in the mouth, mouthfeel, often described as an astringent or drying sensation. It is sometimes mistakenly referred to as a taste, but is formally not one of the basic tastes. Common foods and drinks with astringent character are black tea, red wine, grape skin/seeds, unripe bananas, etc.

Base: Chemically, the opposite of acid. Previously, the term "alkaline" was used for basic substances. A base will react with acids in a neutralization reaction. Basic substances have a pH higher than 7. Few foods are basic, but some examples are soda, egg white in old eggs, and the Nordic speciality "lutefisk" (stockfish, preserved dried whitefish, commonly cod, which is "reconstituted" to resemble fresh fish by soaking in lye solution).

Biochemistry: The branch of natural sciences that studies chemical substances and processes occurring in living organisms.

Carbonate: A negatively charged ion (CO_3^{2-}), which is a part of many naturally occurring salts. Occurs in chemical reactions with carbonic acid (carbonated drinks), carbon dioxide gas, and hydrogen carbonate (bicarbonate), such as the chemical leaveners baking powder and baking soda.

Coagulation: A stronger and irreversible change of proteins compared with denaturation. May be caused by heat, acidic, or basic conditions, or by reaction with enzymes. In coagulation, the long protein molecules react with each other and can form three-dimensional networks in the food. This way, for example a raw egg that is a solution of proteins in water, the networks transform into a gel where water is bound in a network of coagulated proteins. See also, denaturation.

Collagen: Fiber-forming protein, and the most important building block in connective tissues such as cartilage, tendons, and veins/arteries.

Condensation: Phase change of a substance from gas to liquid, such as when water vapor (steam) condenses to liquid water. It is important to note that phase changes, such as condensation and evaporation, are not the same as chemical reactions because during phase change the substance is not changed into new substances. During condensation, much energy is released, so condensation is a heat transfer mechanism in cooking. This comes in addition to the three basic forms of heat transfer; conduction, convection, and radiation.

Conduction/thermal conduction: Heat transfer between a material of higher temperature with another of lower temperature where the two are in direct physical contact. One of the three basic forms of heat transfer (in addition to convection and radiation).

Conductivity: Intrinsic property of a material that represents its ability to conduct heat (or electricity).

Continuous phase: See dispersion, consisting of non-miscible phases forming an even mixture. In a dispersion, the dispersed phase is distributed evenly in the continuous phase. Thus, the continuous phase surrounds small droplets, bubbles, or particles that form the dispersed phase. Examples of continuous phases are the water phase in whipped cream (liquid foam), the water phase in mayonnaise or Hollandaise sauce (oil-in-water emulsions), and the fat in chocolate (dispersion where cocoa- and sugar particles are distributed in cocoa butter).

Convection/thermal convection: Heat transfer in a gas or liquid where material of higher temperature moves to an area of lower temperature. One of the three basic forms of heat transfer (in addition to conduction and radiation).

Cooking loss: Weight loss as a result of cooking (heating) of food, usually meat and fish. Occurs mainly as result of evaporation and drainage of water. Cooking loss is usually given as percentage value.

Denaturation: Mild, reversible or partially reversible, change/modification of structure of proteins. Denaturation can occur when proteins are heated, react with acids or bases, or react with enzymes. Denaturing can also occur by means of mechanical action, such as when whisking egg whites. Denaturation is closely related with coagulation, which is a stronger and irreversible change in proteins for the same reasons. See also, coagulation.

Diffusion: A phenomenon whereby molecules are randomly spread/distributed in a gas or liquid as result of their constant thermal motion. In a gas mixture or solution, substances will gradually be distributed evenly as a consequence of diffusion.

Dispersed phase: See dispersion, consisting of a dispersed phase and a continuous phase. The dispersed phase is the one evenly distributed in the continuous phase. Examples of distributed phase are gas bubbles in a bread (foam), oil droplets in mayonnaise (oil-in-water emulsions), and fruit pulp floating in fruit juice (suspension). See also emulsion, foam and gel.

Dispersion: A mixture of substances that are not miscible where one component is evenly distributed in the other. Examples of dispersions are emulsions (one liquid distributed in another), gels (a liquid distributed in a solid), and foams (gas distributed in either a solid or liquid) are examples. See also, dispersed phase and continuous phase.

Dry matter: The proportion of a food that is not water. For example, the proportion of water in a cucumber is around 96%. The proportion of dry matter is thus 4%, only marginally higher than in seawater (3.5%).

Elastic: A property of a material where it resumes its original shape spontaneously after being deformed (stretched or compressed). Often used in relation with the term plastic. Examples of elastic materials and objects are rubber bands, inflated balls, and a well-developed gluten network.

Emulsion: An even mixture of one liquid into another in which the two liquids are not miscible, for instance oil and water. With the aid of an emulsifier, it is still possible to achieve relatively stable, even mixture: an emulsion. Emulsion is an example of the more general term dispersion. In the kitchen, we have oil-in-water emulsions, where oil is evenly distributed but not dissolved in water (mayonnaise, Hollandaise sauce), or they may be water-in-oil emulsions where water droplets are evenly distributed in oil/fat (butter, margarine).

Enzyme: Biological catalyst, a compound that accelerates biochemical reactions in cells. Chemically, enzymes are a subgroup of proteins. Enzymes require certain conditions to function. They can contribute to food decay but can also be utilized in cooking and food manufacture. The most important factors affecting the enzyme's functions are heat, salinity and pH.

Ester: A category of organic compounds closely related to organic acids (carboxylic acids). The aroma in many fruits and flowers is due to organic esters.

Evaporation: Phase transition from liquid to gas. Although evaporation is most rapid when the liquid boils, it is worth noting that it takes place at almost all temperatures, albeit at different rates (for example, water boils at 100°C but will still evaporate slowly at room temperature).

Falsify: To disprove. Usually used in relation to research, and commonly associated with the philosopher of science Karl Popper, who wrote that hypotheses cannot be proven, only falsified. "Scientifically proven" is thus an unscientific statement. The hypothesis that stands despite all attempts at falsifying it, is often considered the most plausible explanation. When putting forth an explanation, the researcher should therefore strive to make the hypothesis testable, i.e. falsifiable.

Fermentation: Transformation of substances by means of microorganisms. Examples of fermented foods are cheese and other sour milk products, beer and wine, various fermented fish (still-present ancestors of gravlax, e.g. Icelandic hákarl, Norwegian rakfisk, or Swedish surströmming), fermented vegetables such as kimchi and sauerkraut.

Flavor: Complex experience consisting mainly of the sense of taste (in the mouth) and olfaction/smell (in the nose). Other perceptions, experienced through the trigeminal nerves (pain, heat, cooling) and texture are also considered part of, or contributing to, flavor.

Foam: Foams are special cases of dispersions in which a gas is evenly distributed/dispersed in, or surrounded by, a liquid or solid. The gas is always the dispersed phase in a foam. Examples of solid foam are raised breads and cakes, foam rubber, Styrofoam, and meringue. Examples of liquid foam are whipped cream, raw meringue (whipped egg whites), whipped sauces, and cappuccino foam.

Gel: A gel is a special case of dispersion where a liquid is bound, evenly distributed, in a solid. Liquid is the distributed phase and the solid is the continuous phase. Examples of gels are jello, cucumber, cheese, and meat. Jello is a "pure" gel; the others are more complex mixtures but can also be seen as gels.

Gelatinize: Take up liquid to form a gel. An example is starch granules that absorb water.

Gliadin: One of the proteins found in, among others, wheat grain, and together with glutenin forms gluten in pastries, breads, etc. In order for gluten networks to form, glutenin and gliadin must interact with each other. Gluten is a so-called protein complex of the two proteins.

Glucosinolate: Compounds with certain nitrogen and sulfur groups. Present in some plants, they function as defense mechanism against animals and insects. When the plant cells are damaged, they come into contact with enzymes, subsequently liberating pungent tasting/smelling substances. While possibly harmful for humans and animals, they can, in low concentrations, have medicinal properties. They also have desirable flavor characteristics, such as the pungent sensation from mustard, horseradish, wasabi, radish, and certain varieties of lettuce and cabbage.

Gluten: A plastic and elastic network built up of glutenin and gliadin proteins, providing flexibility in doughs with wheat flour. The gluten network makes it possible to produce pasta and raised pastries, breads, etc.

Glutenin: One of the proteins found in, among others, wheat, which together with gliadin forms gluten. See also gliadin and gluten.

Half-life: The time it takes for half the amount of a compound present in a chemical reaction to be consumed. It is one of the measures used to describe the rate of chemical reactions (and for decomposition of radioactive materials).

Heat: A process where heat energy is exchanged between an area or object with higher temperature and another having a lower temperature.

Heat capacity: The property of a material/substance to store thermal energy. The specific heat capacity is the inherent property per quantity of the material, where the unit is J/(kg K). That is, the energy it takes per kg of material to increase its temperature by one Kelvin (or °C).

Hemicellulose: Long-chain molecules (polymers) consisting of between 500 and 15,000 sugar molecules. Hemicellulose is related to cellulose and, together with cellulose and lignin, one of the constituents of the cell walls in plants. Hemicellulose can react with acids and bases. In basic solution it may dissolve, which results in the food to be cooked (or overcooked) quicker.

Hydrophilic: "That which loves water." Property of substances that make them attract water, or are miscible with (soluble in) water. In a chemical sense, this is related to the term polarity; hydrophilic compounds or groups within a compound are usually polar or ionic. Hydrophilic is often set up as contrast to hydrophobic, that is water-repellent properties. Examples of hydrophilic substances are water-soluble substances such as sugar, salt, and alcohol (ethanol).

Hydrophobic: "That which fears water." Properties of substances that make them repel water and, therefore, are not miscible with water. In a chemical sense, this is related to the term polarity; hydrophobic compounds or groups within a compound are usually non-polar. Hydrophobic is often set up as contrast to hydrophilic. Examples of hydrophobic compounds are non-water soluble/miscible substances such as oil and fat. The term is often related to the term lipophilic, "that which loves fat."

Inhibitor: A substance that inhibits chemical or biochemical reactions.

Ion: Atom or molecule with electric charge (positive or negative). Ions form salts. All substances built up by ions will have both positively and negatively charged ions in a way so that the sum of charges balances one another (charge balance).

Kelvin (K): Basic unit for temperature used in scientific contexts. The Kelvin scale has the same relative magnitude as the Celsius scale (one degree Celsius equals one Kelvin, relatively). While 0°C is defined as the temperature for which pure water melts/freezes, 0 K is defined as the lowest possible temperature attainable, absolute zero which is -273.15°C.

Lipophilic: "That which loves fat," substances miscible with fat. Lipophilic substances are usually not miscible or soluble in water. The term is often linked with the term hydrophobic, and used as opposite to hydrophilic.

Maillard reaction(s): A reaction sequence between sugars and amino acids by which color, taste and aromas develop. The Maillard reactions may proceed slowly at low temperatures, but the rate increases radically when the temperature rises to around 140°C. Examples of foods where the Maillard reactions play an important role are bread (brown crust), roast/fried meat (browning of the surface), coffee (during roasting), fudge/caramel, and dulce de leche (due to reaction between lactose and milk proteins during extended boiling).

Micelles: Spherical particles formed when soap molecules, with their hydrophilic and hydrophobic/lipophilic parts, arrange themselves to surround fat droplets, usually in water. The lipophilic parts align toward, and into, the fat droplets while the hydrophilic parts face the surrounding water. This way, the micelles can follow the water when formed in the encounter between fat, soap, and water during dishwashing.

Microorganisms: Common designation for microscopic living organisms that include yeasts, molds and bacteria. Enzymes are chemical substances (proteins) and are therefore not microorganisms.

Molecules: Among the smallest parts of chemical compounds. Molecules consist of two or more atoms.

Mouthfeel: The experience, most often of food or drink, in the mouth that is not directly associated with taste (the "basic tastes," sweet, sour, salty, etc.) or olfaction/smell. Mouthfeel is usually related to texture, viscosity, smoothness, chewiness, crispness, hardness, graininess, juiciness, etc.

MSG/glutamate: Abbreviation of monosodium glutamate. MSG is the sodium salt of an an amino acid called glutamic acid. MSG is found naturally in various foods and raw materials, but is also used in its pure form as an additive (additive 621, in Europe E621). It acts as a flavor enhancer and provides umami flavor.

Multisensory: That which is related to the simultaneous sensation by multiple senses (taste, sight, olfaction/smell, audition/hearing, touch). Related to the concept modality/sensory modality, where a modality is the experience from one of the senses. Eating is a multisensory experience that involves all five senses.

Myosin: Muscle protein involved in contraction of muscle cells. Consequently, it is also a component of meat, and plays an important role in the structure and texture of meat before and during cooking, and when enjoying (or not enjoying) meat.

Non-polar: Used about substances and molecules. Most easily explained as an opposition to polar, in which a molecule has a charge (is an ion) or has an uneven distribution of electrical charge. A non-polar compound will normally not be miscible with or soluble in water. Non-polar compounds are often fat soluble. A good pair for association is water colors (containing polar pigments) and oil paint (containing non-polar pigments).

Octahedron: A geometric structure, a polyhedron, with eight faces. For reference, a cube is also a polyhedron, but with six faces.

Olfaction: Formal term for the sense of smell.

Pectin: Long-chain molecule (polymer) that acts as one of the building blocks in the structure of plants. Metaphorically, one may say that pectin is the glue that holds the cells in the plant together. It is also present in cell walls, and if the cellulose molecules are seen as the reinforcing bars in the cell wall, the pectin can be seen as the cement. Pectin is abundant in the peel from apples and citrus fruits. It is also found in berries but is broken down during ripening. Pectin is used in food and cooking as a gelling agent.

pH: A quantitative measure of how acidic or basic/alkaline a water solution is. A neutral solution has a pH of 7, an acidic solution has a pH below 7 and a base has a pH above 7. The pH scale ranges from 0-14, although it is possible to have solutions with a negative pH and pH greater than 14.

Phase/State of matter: See state of matter.

Plastic: A property described as a material being irreversibly deformed upon pressure. Often seen in relation with the term elastic. An example of a plastic material that is not elastic is clay or play dough. A good bread dough is both plastic and elastic.

Polarity: A property of molecules that describes how charge is distributed as a result of the distribution of electrons and electrical charges within the molecule. For example, water is a polar molecule: the area around the oxygen atom is partially negatively charged and the area around the hydrogen atoms is partially positively charged. Ions and salts are by nature highly polar. By contrast, charge and electrons in non-polar molecules are more evenly distributed. Polarity is a fundamentally important notion in chemistry, and the foundation (among others) for substances being hydrophilic and hydrophobic.

Precipitation: When a substance is converted from water-soluble to insoluble, or when two water-soluble substances react to form an insoluble product. Examples are the coagulation of proteins (a substance is converted from water soluble to insoluble) or the reaction between tannins and proteins to give a precipitate during cooking of stock (reaction between two soluble substances to give a third, insoluble one).

Protease: Common designation for enzymes that break down, or cleave, proteins.

Protein: An organic compound consisting of a long chain of amino acids. Proteins can also be composed of more than one chain that are more loosely bound to each other. Proteins, together with fats and carbohydrates, are among the basic nutrients.

Radiation: See thermal radiation.

State of matter/Phase: A concept describing the physical state of substances. Common states of matter, or phases, are gas, liquid, and solid. Some other states of matter also exist, like plasma and supercritical phase, but these are not commonly encountered in food (although the former can actually be achieved in a an ordinary microwave oven and the latter is used in food industry when supercritical CO_2 is used to remove caffeine from coffee to produce decaffeinated coffee).

Solution/dissolve: Mixture of substances in a liquid phase. The dissolved substance may be a gas, another liquid, or a solid. Examples are sugar or salt dissolved in water, vinegar in water, color or flavorings in oil, carbon dioxide in water (e.g. sparkling). If the substance is completely dissolved, the solution will be completely transparent, but it may still have color. A cloudy or misty solution indicates that it contains some undissolved material, but it can still be a solution since some substances may be dissolved while others are not. To dissolve is not the same as to melt (which is the phase transition between solid and liquid).

Surface tension: A physical surface phenomenon that causes a water surface to behave like a flexible membrane. It is the reason why a needle can float on water or insects may walk on water. Surface tension is also the phenomenon behind capillary forces, that plants (including tall trees) can draw up water through thin capillaries so that the water reaches the whole plant. The reason for the surface tension is weak chemical bonds between water molecules, and between water molecules and other surfaces.

Suspension: An even mixture of two non-miscible substances in which the distributed phase is a solid and the continuous phase is a liquid. One example is orange juice (as opposed to transparent apple juice, which is a solution). Suspension is a special case of dispersions. There also exist solid–solid suspensions, where a solid is distributed in another solid, such as solid chocolate where cocoa and sugar particles are dispersed in solid cocoa fat.

Tannin: Chemical substances (polyphenols) in plants. Tannins react with proteins in precipitation reactions (forming precipitates), during cooking as well as in the mouth. In food and drinks, tannins can give the mouthfeel experience called astringency by reacting and binding proteins in the saliva. This is the dry feeling experienced when drinking certain red wines, strong tea, or biting into a grape stone or skin. Also, tannins may react with proteins during cooking, e.g. when heating meat stock with red wine. In plants, the presence of tannins functions as a protection, as it prevents the plant from being eaten by certain animals and insects.

Taste: One of the five senses (the others are olfaction/smell, sight, audition/hearing, and touch). Taste is, as olfaction, among the chemical senses where the signals to the brain occur as result of chemical substances interacting with taste receptors on the tongue and elsewhere in the mouth.

Thermal energy: The energy of a substance, system or object possesses due to its temperature.

Thermal radiation: In cooking contexts, electromagnetic radiation, usually in the form of infrared radiation, is one of the three basic forms of heat transfer (in addition to conduction and convection). Heat can also be transferred via microwave radiation, which has a somewhat different mechanism from infrared radiation.

Umami: One of the "basic tastes," together with sweet, salty, bitter, and sour (there are probably a number of other such tastes as well). Umami is often referred to as the fifth taste and can be described as rich and brothy. MSG, monosodium glutamate, is among the substances with a distinctive taste of umami. Umami can be experienced as a naturally present taste in many foods, or can be introduced or enhanced by addition of e.g. MSG.

Viscosity: A property of a substance (usually liquid or gas) that describes its ability to withstand flow. It can be associated with how "thick" a fluid is or how easily it flows. Low viscosity means that the substance flows more easily (gases have very low viscosity, water has low viscosity among liquids) while high viscosity means that it flows slowly (honey has high viscosity compared to water).

Water Activity: A numerical value between 0 and 1 that indicates the amount of water in a food that is is not chemically bound to any ingredient, and thus available to microorganisms. For example, the use of salt and sugar for preserving food is based on the fact that these ingredients result in reduced water activity, and thus making the food a "dry desert" for microorganisms that will consequently not thrive and multiply.

Bibliography

1 The Real Thing? The Chemistry and Psychology of an Apple Pie

Food Timeline. FAQs: Pie & pastry. http://www.foodtimeline.org/foodpies.html.

Liberty Letters. Confederate civil war recipes. http://libertyletters.com/resources/civil-war/confederate-civil-war-recipies.php.

Mielby, L. H. & Bom Frøst, M. (2010). Expectations and surprise in a molecular gastronomic meal. *Food Quality and Preference, 21*(2), 213–224. https://doi.org/10.1016/j.foodqual.2009.09.005.

Song, H. J. & Schwarz, N. (2009). If it's difficult to pronounce, it must be risky. *Psychological Science, 20*(2), 135-138. https://doi.org/10.1111/j.1467-9280.2009.02267.x.

2 Boiling an Egg from the Inside Out

ChefSteps (s.a.). *The Egg Calculator.* Retrieved 14th January 2019 from https://www.chefsteps.com/activities/the-egg-calculator.

Henry, A. (2011, August 23). *The Egg Cooking Calculator Uses Math to Help You Boil the Perfect Egg.* Retrieved 14th January 2019 from http://lifehacker.com/5833410/the-egg-cooking-calculator-uses-math-to-help-you-boil-the-perfect-egg.

Stølen, S. & Vedde, J. (2010, December 1). *Kunsten å koke et egg (The Art of Cooking an Egg).* Retrieved 14th January 2019 from http://www.mn.uio.no/kjemi/tjenester/kunnskap/egg.

This, H. (2006). *Molecular Gastronomy. Exploring the Science of Flavor.* New York: Columbia University Press, 29–31.

Vega, C. & Mercadé-Prieto, R. (2011). Culinary biophysics: On the nature of the 6X°C egg. *Food Biophysics,* 6(1), 152–159. https://doi.org/10.1007/s11483-010-9200-1.

3 In the Beginning There Was an Egg

Belitz, H.-D., Grosch, W. & Schieberle, P. (2009). *Food Chemistry*. 4th edition. Berlin: Springer, 515.

Söderberg, J (2013). *Functional properties of legume proteins compared to egg proteins and their potential as egg replacers in vegan food*. Second cycle, A2E. Doctoral thesis. Uppsala: SLU, Department of Food Science. http://stud.epsilon.slu.se/6240/.

Vega, C., & Sanghvi, A. (2012). Cooking literacy: Meringues as culinary scaffoldings. *Food Biophysics*, 7(2), 103-113. https://doi.org/10.1007/s11483-011-9247-7.

Vegetarian times (2004): Chewy meringue cookies. https://www.vegetariantimes.com/recipes/chewy-meringue-cookies.

Wikipedia: Aquafaba. Retrieved 14th January 2019 from https://en.wikipedia.org/wiki/Aquafaba#Etymology.

4 Mussels on Acid

Belitz, H.-D., Grosch, W. & Schieberle, P. (2009). *Food Chemistry*. 4th edition. Berlin: Springer, 56.

Fink, A. L., Calciano, L. J., Goto, Y., Kurotsu, T. & Palleros, D. R. (1994). Classification of acid denaturation of proteins: Intermediates and unfolded states. *Biochemistry*, 33(41), 12504–12511. https://doi.org/10.1021/bi00207a018.

5 Sausage Engineering

Finnish Meat Information Association (s.a.). *Lihankulutus Suomessa (Meat consumption in Finland)*. Retrieved 12th March 2019 from https://www.lihatiedotus.fi/tilastotietoa/lihankulutus-suomessa.html.

National Hot Dog and Sausage Council (2016). *Consumption stats - From the National Hot Dog and Sausage Council*. Retrieved 14th January 2019 from http://www.hot-dog.org/media/consumption-stats.

Savic, I. V. & Food and Agriculture Organization of United Nations (FAO) (1985). *Small-scale sausage production*. Retrieved 14th January 2019 from http://www.fao.org/docrep/003/x6556e/x6556e00.htm.

6 Mustard - Fiercely Yours

Ali, H. M., El-Gizawy, A. M., El-Bassiouny, R. E. I. & Saleh, M. A. (2015). Browning inhibition mechanisms by cysteine, ascorbic acid and citric acid, and identifying PPO-catechol-cysteine reaction products. *Journal of Food Science and Technology*, 52(6), 3651–3659. https://doi.org/10.1007/s13197-014-1437-0.

Botti, M. G., Taylor, M. G. & Botting, N. P. (1995). Studies on the mechanism of myrosinase. Investigation of the effect of glycosyl acceptors on enzyme activity. *The Journal of Biological Chemistry*, 270, 20530–20535. https://doi.org/10.1074/jbc.270.35.20530.

Raghavan Uhl, S. (2000). *Handbook of Spices, Seasonings, and Flavorings*. CRC Press, 136.

7 The Potatoes that Refused to Become Tender

Andersson, A., Gekas, V., Lind, I., Oliveira, F. & Oste, R. (1994). Effect of preheating on potato texture. *Critical Reviews in Food Science and Nutrition*, 34(3), 229–51.

Anthon, G. E. & Barrett, D. M. (2006). Characterization of the temperature activation of pectin methylesterase in green beans and tomatoes. *Journal of Agricultural and Food Chemistry*, 54(1), 204–211. https://doi.org/10.1021/jf051877q.

8 Gravlax and a Pinch of Salt Chemistry

Davidson, A. & Jaine, T. (2014). *The Oxford Companion to Food*. Oxford: Oxford University Press, 352.

Heuch, H. & Riddervold, A. (2006). *Rakefisk: "- en vidunderlig spise"*. Oslo: Damm.

Wang, D., Tang, J. & Correia, L. R. (2000). Salt diffusivities and salt diffusion in farmed Atlantic salmon muscle as influenced by rigor mortis. *Journal of Food Engineering*, 43(2), 115–123. https://doi.org/10.1016/S0260-8774(99)00140-5.

9 The Art of Heating

Baldwin, D. (2010). *Sous Vide for the Home Cook*. Paradox Press LLC. See also http://www.douglasbaldwin.com/sous-vide.html.

Fooladi, E., Isnes, A. (2009, August 18). *Varme og matlaging (Heat and Cooking)*. National Centre for Science Education in Norway. Retrieved 14th January 2019 from http://www.naturfag.no/artikkel/vis.html?tid=1227355.

Myhrvold, N., Young, C. & Bilet, M. (2011). *Modernist Cuisine. The art and science of cooking*. (Volume 1, Chapter 5). Washington: The Cooking Lab, 206–291.

Wolke, R. (2002). *What Einstein Told His Cook. Kitchen science explained*. New York: W. W. Norton & Company, 190–194.

10 Cold Skillet - Juicy Fish

Myhrvold, N., Young, C. & Bilet, M. (2011). *Modernist Cuisine. The art and science of cooking*. (Volume 3, Chapter 11). Washington: The Cooking Lab, 70–115.

11 Tempering the Meat Before Cooking, ...Or Maybe Not

Roe, K., Breiman, L. & Stephens, M. (2013). *"Cook my meat calculator"*. Massachusetts Institute of Technology. Retrieved 14th January 2019 from http://up.csail.mit.edu/science-of-cooking/home-screen.html.

Souza, D. & America's Test Kitchen. (2014, July 30) *Science: Make the best Steaks by Cooking Frozen Meat (No Thawing!)*. Retrieved 14th January 2019 from https://youtu.be/uLWsEg1LmaE.

12 Cooking Pits - Prehistoric Cooking Meets Science

Dunster, E. S. (1870). The self-acting Norwegian cooking apparatus. *The New York Medical Journal, vol. 10,* 108–111. https://books.google.it/books?id=CzcCAAAAYAAJ&pg=PA109.

Eat this podcast (2016). The haybox through history. http://www.eatthispodcast.com/the-haybox-through-history.

Fooladi, E. (s.a.). *Mat i kokegropa (Food in Cooking Pit)*. Retrieved 14th January 2019 from http://www.natursekken.no/uopplegg/vis.html?tid=1188274.

Iversen, F. & Petersson, H. (Eds.) (2016). *The agrarian life of the North 2000 BC – AD 1000 : Studies in Rural Settlement and Farming in Norway.* Kristiansand: Portal. https://press.nordicopenaccess.no/index.php/noasp/catalog/book/13.

Narmo, L. E. (1996). Kokekameratene på Leikvin - Kult og kokegroper. *Viking* , vol. 59, 79–100. http://www.arkeologi.no/viking/tidligere.

Wandsnider, L. (1997). The roasted and the boiled: Food composition and heat treatment with special emphasis on pit-hearth cooking. *Journal of Anthropological Archaeology*, 16(1), 1–48.

13 White Wine in Cooking Should Always Be Dry, Right?

Edison, T. (2019, January 1). *9 Best White Wines for Cooking*. Retrieved 14th January 2019 from http://www.wineturtle.com/best-dry-white-wine-for-cooking.

Henderson, J. (2018, November 21). *The 5 Best White Wines for Cooking*. Retrieved 14th January 2019 from http://www.thekitchn.com/5-white-wines-for-cooking-202573.

Robuchon, J. (2001). *Larousse gastronomique. The World's Greatest Culinary Encyclopedia*. Lang, J. H. (Ed.). London: Hamlyn. 1298–1301.

Rognså, G. H. (2014). *Emulsions from a culinary perspective - The case of Hollandaise sauce and its derivatives*. Doctoral thesis. Department of Food Science, University of Copenhagen. http://research.ku.dk/search/?pure=en/publications/emulsions-from-a-culinary-perspective(4ed5e224-390d-47bf-9f3e-b89b52ea6e4d).html.

Sivertsvik, M., Rognså, G. H., Rathe, M., Agerlin Petersen, M., Misje, K.-E., Hersleth, M. & Risbo, J. (2017). From wine to wine reduction: Sensory and chemical aspects. *International Journal of Gastronomy and Food Science*, 9, 62–74. https://doi.org/10.1016/j.ijgfs.2017.06.006.

Sivertsvik, M., Rognså, G. H., Rathe, M., Agerlin Petersen, M., Misje, K.-E., Brüggemann, D. A., Hersleth, M. & Risbo, J. (2017). From wine to hollandaise sauce: Does the nature of the wine or wine reduction influence sensory attributes? *International Journal of Gastronomy and Food Science*, 9, 75–87. https://doi.org/10.1016/j.ijgfs.2017.06.003.

14 Taking Stock of Broth

Snitkjær P. (2010). *Investigations of meat stock: From a molecular gastronomy perspective.* Doctoral thesis. Department of Food Science, University of Copenhagen. http://static-curis.ku.dk/portal/files/32448394/PHD.0111.pdf.

This, H. (2009). *Building a Meal. From Molecular Gastronomy to Culinary Constructivism*. Translated by Malcolm DeBevoise. New York: Columbia University Press, 39–54.

15 Salt Shapes the Pasta

Tuhumury, H.C.D., Small, D.M. & Day, L. (2014). The effect of sodium chloride on gluten network formation and rheology. *Journal of Cereal Science, 60*(1), 229–237. https://doi.org/10.1016/j.jcs.2014.03.004.

16 You Can See It, Hear It, Taste It and Feel It - What Is It? It Is the Taste of Food!

Abrahams, M. (2004, March 8). *The sound of flavor: A review of the research.* Retrieved 14th January 2019 from http://www.improbable.com/2015/03/08/the-sound-of-food-a-review-of-the-research.

Hopia, A. & Ihanus, S. (Eds.) (2014, May 5). *5D Cookbook*. Retrieved 14th January 2019 from http://www.5DCookBook.fi.

Holt-Hansen, K. (1968). Taste and pitch. *Perceptual and Motor Skills*, 27(1), 517–538. https://doi.org/10.2466/pms.1968.27.1.59.

Morrot, G., Brochet, F. & Dubourdieu, D. (2001). The colour of odors. Brain and Language, 79(2), 309–320. https://doi.org/10.1006/brln.2001.2493.

Spence, C., Richards, L., Kjellin, E., Huhnt, A.-M., Daskal, V. D., Scheybeler, A., Velasco, C. & Deroy, O. (2013). Looking for crossmodal correspondences between classical music and fine wine. *Flavour, 2*(1), 29. https://doi.org/10.1186/2044-7248-2-29.

Zampini, M. & Spence, C. (2004). The role of auditory cues in modulating the perceived crispness and staleness of potato chips. *Journal of Sensory Studies*, 19(5), 347–363. https://doi.org/10.1111/j.1745-459x.2004.080403.x.

17 Can You Cut Down on the Salt Without Affecting the Taste?

Cho, H., Kim, S. M., Jeong, S. S. & Kim, S. B. (2016). Comparison of salt taste thresholds and salt usage behaviors between adults in Myanmar and Korea. *Asia Pacific Journal of Clinical Nutrition 25*(4), 879–884. http://doi.org/10.6133/apjcn.092015.32.

Elmer, P. J. (1988). *The effect of dietary sodium reduction and potassium chloride supplementation on sodium chloride taste perceptions in mild hypertensives.* Unpublished doctoral thesis. University of Minnesota. Referenced in: IOM (Institute of Medicine), Henney, J. E., Taylor C. L. & Boon C. S., (Eds.) (2010). *Strategies to Reduce Sodium Intake in the United States.* Washington, DC: The National Academies Press. https://www.ncbi.nlm.nih.gov/books/NBK50958.

European Commission (s.a.). *Salt campaign.* Retrieved 14th January 2019 from https://ec.europa.eu/health/nutrition_physical_activity/high_level_group/nutrition_salt_en.

Gillette, M. (1985). Flavor effects of sodium chloride. *Food Technology*, 39, 47–56. Referenced in Liem, D. Gie., Miremadi F. & Keast R. S. J. (2011). Reducing Sodium in Foods: The Effect on Flavor. Nutrients, 3(6), 694-711. http://doi.org/10.3390/nu3060694.

Henney, J. E., Taylor, C.L. & Boon, C. S. (Eds.) (2010). *Strategies to Reduce Sodium Intake in the United States.* Institute of Medicine (US) Committee on Strategies to Reduce Sodium Intake. https://www.ncbi.nlm.nih.gov/books/NBK50956.

Lawrence, G., Salles, C., Septier, C., Busch, J. & Thomas-Danguin, T. (2009). Odour–taste interactions: A way to enhance saltiness in low-salt content solutions. *Food Quality and Preference, 20*(3), 241–248. http://doi.org/10.1016/j.foodqual.2008.10.004.

Lersch, M. (2007, May 1). *Practical molecular gastronomy, part 5.* Retrieved 14th January 2019 from https://blog.khymos.org/2007/05/01/practical-molecular-gastronomy-part-5.

Nasri, N., Septier, C., Beno, N., Salles, C. & Thomas-Danguin, T. (2013). Enhancing salty taste through odour–taste–taste interactions: Influence of odour intensity and salty tastants' nature. *Food Quality and Preference, 28*(1), 134–140. http://doi.org/10.1016/j.foodqual.2012.07.004.

The Norwegian Directorate of eHealth (2017, June 21). *Hvordan spise mindre* *salt? (How to Eat Less Salt?)* Retrieved 14th January 2019 from https://helsenorge.no/kosthold-og-ernaring/kostrad/salt/hvordan-spise-mindre-salt.

Umami Information Center (2017). *Umami Rich Ingredients.* Retrieved 14th January 2019 from https://www.umamiinfo.com/richfood.

US Food and Drug Administration (2018, August 2). *Sodium Reduction.* Retrieved 14th January 2019 from https://www.fda.gov/food/ingredientspackaginglabeling/foodadditivesingredients/ucm253316.htm.

World Health Organization (2016, June 30). *Salt reduction.* Retrieved 14th January 2019 from http://www.who.int/news-room/fact-sheets/detail/salt-reduction.

World Health Organization Europe (2014, September 26). *Reducing salt consumption.* Retrieved 14th January 2019 from http://www.euro.who.int/en/health-topics/disease-prevention/nutrition/news/news/2014/09/reducing-salt-consumption.

18 Physics Takes the Cake

Barham, P. (2001). *The science of cooking.* Berlin: Springer, 151–157.

19 The Great Norwegian Porridge Feud

Brekke, L., Fet, M. & Fet, J. (2015). Personal communications, and mail correspondence August 2015.

Brown, C. (2010). *Broths to Bannocks: A History of Cooking in Scotland from 1690 to the Present Day.* Lanark: Waverley Books Ltd.

Haggrén, G., Halinen, P., Lavento, M., Raninen, S. & Wessman, A. (2015). Muinaisuutemme jäljet. Helsinki: Gaudeamus.

Kolmonen, J. & Sommar, H. (2017). Cooking potato casserole (video). YLE - Finnish Broadcasting Company. https://yle.fi/aihe/artikkeli/2009/12/17/imelletyn-perunalaatikon-valmistus.

Riddervold, A. & Ropeid, A. (1984). Popular diet in Norway and natural science during the 19th century. The porridge feud 1864–66. *Ethnologia Scandinavica,* 48–65.

Ropeid, A. (1972). Fornuftigt madstel. Oslo: By og bygd - Yearbook of the Norwegian Museum of Cultural History 1971–72, 77–112.

Sanden, J. (2006). Tradisjonsmat i Romsdal. Med oppskrifter. In J. Sanden (Ed.), Yearbook of Romsdal Museum (Vol. 2006, 11–57). Molde: Romsdal Museum. Retrieved 14th January 2019 from https://www.romsdalsmuseet.no/no/kunnskap/arboker.

20 Doing the Dishes

No references.

Index

C

D

E

F

G

H

I

N

O

P

R

S

T

U

V

W

Z